I dedicate this book to my parents,
with thanks for always believing my time would come.

CONTENTS

I

THE FISH MERCHANT'S DAUGHTER

It was the golden summer of 1903, a glorious era of British history that has become known as the Edwardian Age, when the blazing red of the British Empire spanned the globe, and its malevolent influence seemed omnipotent, almost invincible, that Amy Johnson was born on 1st July in the family home at 154 St Georges' Road, Hull.

It was a long, tree-lined street of small terraced houses, with bay windows, beyond which lay the sprawling St Andrew's Dock, in an area that Amy remembered for the pervading smell of fish.

Hull was a thriving port city. Strutting out into the Humber Estuary, it basked in a strong sense of civic pride and the discovery of rich hunting grounds close to the Dogger Bank attracted fishermen from as far south as Ramsgate and Brixham, enabling Hull to establish its status as the third largest port in the United Kingdom.

Amy's father Will Johnson acquired the house in the autumn of 1902 shortly after his marriage to Amy Hodge, who was always known within the family by her childhood name of 'Ciss.' Will was a market salesman in the firm Andrew Johnson

Knutzdon, or 'AJK' as it was generally known within the trade, had been established by his father and a Norwegian partner who died a few years later.

Anders Jorgensen, was the sixteen-year-old son of a well-digger from a small village near Assens, when he left his Danish homeland to cross the North Sea. After eight grueling years, during which he served his apprenticeship aboard a fishing smack, climbing up the ranks to be skipper, he decided to settle in Hull. Taking British citizenship and anglicising his name to Andrew Johnson.

Will was the eldest son from Andrew's marriage to Mary Ann Holmes. He inherited his father's shrewd business brain and his apparent lust for adventure.

In 1898 twenty-one-year-old Will set off with three friends for the Klondike to seek his fortune in the gold fields of Atlin, British Columbia.

It was a makeshift, overcrowded town consisting of row upon row of hastily pitched tents, in a freezing, desolate place, where the harsh laws of nature held sway and a man's endurance was tested to it's limits, but his dreams and aspirations turned swiftly to despair.

Will returned home after two years with just four gold nuggets to show for his efforts. His gold prospecting days were over and although mildly disappointed by this setback, he succumbed to the routine of his job, imbued with a fresh determination to combine his flair and ambition, to making a success of the herring trade.

While the fortunes of the Johnson family were on an upward curve, the contrast with those of the Hodge family could not have been greater. Amy's maternal great-grandfather held public office in the city during the mid-19th century, eventually becoming Mayor, and bestowing upon Hull a new town hall. He also acquired a great house befitting the family's exalted status in the city. Newington Hall on Anlaby Road with its Corinthian pillars and grand entrance hall had once been the home of a local grocer. The exorbitant cost of its maintenance devoured the family fortune, until it was abandoned for a more modest property on Coltman Street. Newington Hall remained, however, as a painful reminder of better times, until its demolition in 1908.

After William Hodge's death, his eldest son developed a taste for extravagant living. He travelled abroad frequently, abandoning the care of their children to his long-suffering wife, and although he made a few speculative investments, including a brief foray into timber merchanting, he lacked his father's business flair and stamina.

His investments never quite realised their potential, and the linseed mills, which were once the staple of the Hodge family fortune were eventually sold off. 'Ciss' Hodge with her fragile, artistic temperament, was deeply affected by this dramatic fall in the family's circumstances, and was acutely aware that through her engagement to Will Johnson, she was taking a step down the social ladder by marrying into a family considered less distinguished than her own.

Amy was just five months old and protected in the blissful cocoon of infancy when the historic event that would ultimately change the course of her adult life occurred one cold winter day 3,000 miles away, on a stretch of North Carolina beach. In the curiously named *Kitty Hawk*, two brothers from Ohio, Wilbur and Orville Wright finally fulfilled man's long-held ambition to fly; by keeping an heavier-than-air machine in the sky for just twelve seconds, as an icy Atlantic wind blew across the sand dunes on Kill Devil Hill.

It was their third attempt of the morning and their patience was beginning to fray, but the world took little notice of what appeared to be just another foolish attempt at man-powered flight. However incredulous it might seem, from this humble beginning the Age of Aviation was born. Within the decade, a

Frenchman named Louis Bleriot had flown across the Channel and a decade on from that John Alcock and Arthur Whitton Brown succeeded in an audacious flight across the Atlantic.

Within the first two decades the fledging world of aviation beguiled and fascinated as it seeped into the consciences of people across the globe. One brave pilot after another stepped forward to attempt awe-inspiring flights, often risking life and limb in the pursuit of personal glory, and financial reward and occasionally in a genuine attempt to advance the cause of aviation. Whatever the personal motive, there was one common objective: *the Conquest of the Skies.*

Aviation remained however a solidly male dominion, one of the last bastions of chauvinism. Flying lessons were expensive, and therefore beyond the pocket of the average man, while, with the exception of the great American *aviatrix* Harriet Quimby, who in April 1912 became the first woman to fly solo across the Channel, the few female pilots who emerged came from the upper echelons of society, such as the Duchess of Bedford and Lady Mary Bailey who had the time and a wealthy husband, willing to indulge them in an expensive and dangerous new hobby.

In those pioneering days it seemed impossible that any woman could break down the wall of prejudice, which had built up around aviation and make a significant contribution to its future development, least of all a fish merchant's daughter from Hull.

For the first eighteen months of her life, Amy became the centre of her parents' world and enjoyed the luxury afforded to the first-born child of undivided attention. That was until November 1904 when 'Ciss' Johnson gave birth to a second daughter; a sister for Amy, christened Irene, was welcomed into the solid red-brick Victorian home on St George's Road. Although a proud and doting father to his two small daughters, Will Johnson longed for a son, a male heir who would join him in the family business at 'AJK' and work his way up to eventually take over the reins in much the same way that Will was steadily doing for his father Andrew.

So Will Johnson welcomed the birth of his next two children, with the same feelings of hope and expectation with which he had awaited the arrival of Amy and Irene, but his wish for a son, was never fulfilled.

'Ciss' Johnson adapted quickly to life as a young wife and mother, becoming a proud and dedicated home-maker, executing her domestic chores with a kind of evangelical zeal. She liked singing her favourite hymns, in a high melodious voice, and though she still suffered with her nerves, 'Ciss' found her pleasure; playing the piano and chapel organ. During the Hodge family's affluent times while ensconced in the splendour of Newington Hall, William Hodge had demonstrated considerable largesse, in funding chapels that were committed to Primitive Methodism.

The chapel was always an integral part of 'Ciss' Johnson's life. Her adherence to the *Edwardian Sabbath* was absolute, and her unshakeable devotion to the Methodist beliefs in which she had been raised, would later lead to a serious breach with her eldest daughter.

When Amy was six, the family moved a few streets east, to the Boulevard, and then in 1910 to 48, Alliance Avenue. It was here that Mollie was born in 1912. Each move took the family further away from St Andrew's Dock, and the all-pervading smell of fish. Will also cemented his growing affluence and reputation, as a businessman of some acumen, joining the Rotary Club and becoming a Freemason. Yet he retained a common touch, and as a supporter of Hull City Football Club, he enjoyed the jostle and banter of the terraces on a Saturday afternoon.

In 1914 Andrew Johnson finally retired officially from 'AJK' leaving Will to fulfil the role, that he had effectively held for some years as the titular head of the family business.

In August war was declared in Europe after a long intense summer of frenzied diplomatic brinkmanship, which had followed the assassination in June of Archduke Ferdinand of Austria at Sarajevo. As the unfound optimism, bordering on hyperbole, that the war would be over by that first Christmas 1914 vanished amid the mud, rain and the unimaginable carnage of Flanders, Will Johnson became deeply concerned about the effect that a long ensuing conflict could have on his commercial interests. The contacts he had established in Scandinavia benefited from Norway's neutrality, but as the war dragged on without an end in sight, the strain and worry began to have a devastating effect on Will's health. He contracted a severe skin condition called erysipelas which rendered him a virtual housebound invalid. A nurse moved in to care for him, and the loss of status, in having the girls see their strong, dependable father, confined to his bed, with his face swollen, red and blotchy, had a deep affect on him. Meanwhile 'Ciss' with her tenuous nerves stretched still further vacillated between her combined roles, as the loving and concerned wife to Will and as a calm, soothing influence on three impressionable girls.

Amy had celebrated her eleventh birthday a month before the hostilities began, and she viewed the conflict from an entirely different perspective: blessed, as she was with the innocence of childhood the war had had a strangely positive impact on her development, opening up new vistas of experience, lifting her out of her typically dull, routine, life.

The Zeppelin raids over Hull, although very frightening, were also illuminating, especially witnessing people's reactions, whenever the sirens sounded and what action they took to protect themselves from attack. There were no official air-raid shelters and while some took advice to take cover beneath the stairs, others headed for open ground, believing it was safer, and then jumped into ditches whenever the huge airborne monsters loomed menacingly overhead.

As a port city, and a vital link in the supply chain, Hull was an inevitable target for Zeppelin raids, and in one attack, when two bombs dropped, twenty-five people were killed and many more injured inciting accusations that the government was not doing enough to protect people from danger.

Amy viewed it as a treat, to be allowed to stay up long after her official bedtime whenever the air-raid sirens sounded, but when Hamlyn Petrie, the husband of Will's sister Evelyn, was killed in France, any lingering romantic illusions that the girls may have harboured about the true nature of war were cruelly and irretrievably shattered as the harsh realities of the conflict were brought home to them in tangible human terms.

In late August 1915 'Ciss' took Amy and Irene to be enrolled at Boulevard Municipal Secondary School, and although Amy was put in a class with children younger than herself due to an administrative error at the Eversliegh School, which the girls had left in June shortly before Amy's twelfth birthday, she quickly excelled at the Boulevard, jumping to the top of the class, and exerting some authority over her younger classmates.

The fastidious and old fashioned principal, Frederick De Velling, a career academic and strict disciplinarian, was to have a profound, and largely positive influence on Amy's developing character. Although her obvious academic abilities delighted him, her taste for rebellion, and challenges to school authority would invariably exasperate him throughout the seven long years that she spent within the confines of his education establishment.

Amy's appetite for rebellion was to be satisfied in various ways during her time at the Boulevard, and one such incident, which was to gain some notoriety, was over a proposed swimming gala which she wanted the school to organise. Swimming and hockey were two of Amy's main interests, but De Velling didn't consider swimming to be particularly ladylike, and he refused. Unperturbed, Amy and Irene cajoled their father into hiring the

Beverley Road public swimming-baths, which they frequently used, for the afternoon so that the gala could go ahead in spite of De Velling's objections.

She also persuaded her French tutor, Mary Sheppard to adjudicate at the gala and present the prizes. Although a very friendly and approachable teacher, who was to inspire such respect, and affection from her pupils that she was visited by so many of her former charges long into her retirement, she was naturally reluctant to risk flouting De Velling's authority *so* openly. Mary Sheppard eventually agreed as long as the gala attracted no publicity, and so Amy enjoyed a minor triumph over De Velling, but some of her other attempts at rebellion would not be so successful.

In the lower forms the male and female pupils were kept apart so when Amy was seen casually walking in Pearson Park with a male pupil while dressed in her school uniform, she was reported by a member of the teaching staff, and had to face the wrath of the headmistress, Edith Clews, as a result.

The Armistice in November 1918 came as a great relief to Will Johnson, albeit tinged with some sadness due to the loss of his brother Bert. The end of hostilities meant a return to some semblance of normality. Business had picked up and Will had regained his confidence; there was a spring in his step once again.

He planned his last and most ambitious house move in Hull – moving the family to Park Avenue, a prosperous area on the northern reaches of the city, where his neighbours included solicitors and an American diplomat.

The onset of puberty made Amy introverted. She felt self-conscious, and slightly 'gauche', caused perhaps by her gaining weight, and she sought more time on her own. The urge for solitude was exacerbated in the spring of 1919 when 'Ciss' Johnson gave birth to a fourth child. It was an unplanned and last vain attempt to gain a male heir. Amy was jealous, and slightly embarrassed, by the new arrival, who was christened Betty. There was a difference of sixteen years between the eldest and youngest Johnson girls, and as her parents' time and attention was devoted to their new-born daughter, Amy yearned for any means of escape. The gift of a bicycle gave her the freedom that she so craved, and she exploited it voraciously.

As the austerity and gloom of post-war England began to fade, and the patriotic songs that were designed to rally people around the flag gave way to the upbeat flamboyant sound of Ragtime and Jazz. The new decade brought with it a renewed sense of hope for better things to come.

The Roaring Twenties or the Jazz Age as the decade also became known, spawned among other novelties, the Flapper, the easy-going short-haired girl who would dance the night away in speakeasies across America. The Charleston represented more than just the latest dance craze. It became the cultural phenomenon that swept across the Atlantic and when eighteen-year-old Amy, herself a slavish follower of the latest fashion or fad, decided to try 'bobbing' her hair by hacking away two plaits, along with the maroon school ribbon that she so detested, Will was horrified, but he consoled himself with the fact that Amy's hair would grow back in time for her graduating from the Boulevard.

Amy should have graduated in the summer of 1921 shortly before her eighteenth birthday, but the administrative error in junior school about her age had dogged her progress throughout her years at the Boulevard. So Amy had to suffer yet another year as the oldest girl in her class, repeating the lessons which had failed to inspire her the first time around.

Will and 'Ciss' Johnson were deeply concerned about the career path their eldest daughter would pursue after graduating highschool. Amy was intelligent, but so easily distracted. A solution of sorts came via Mr De Velling, an academic who had devoted his working life to education and for him there was no greater vocation. Now he had the example of his daughter to extol to his pupils. Elizabeth De Velling was reading Modern Languages, with a view to pursuing a career in teaching, and De Velling basked in her reflective glory unashamedly. The Board of Education was also eager to encourage young people into teaching by rewarding generous grants towards the cost of tuition to anyone of sufficient potential willing to pursue a career in academia. It was seen as a positive move to produce a crop of new younger teachers, away from the middle-aged spinster types who with only the 'right' background and minimal training had taught for years in establishments such as the Eversliegh School.

In 1921 Will took the family to Bridlington, to escape the stifling heat of the city in high summer. 'Brid' as it was affectionately known had become increasingly popular its proximity to Hull made it easily accessible. The summer was spent visiting aunts and uncles and the Johnson grandparents. Will's sister, Evelyn Petrie, had, together with her son John, moved in with her parents due in some part to her reduced circumstances as a war widow. Evelyn loved playing tennis on long summer afternoons and she encouraged Amy to take part. She also introduced her 'gauche' young niece to a very handsome Swiss in his mid-twenties, Hans Arregger, who spoke four languages.

He had moved from Amsterdam to work at the Swiss Consulate in Hull. He was short and stocky, fair-haired with piercing gray eyes, and he possessed a worldly sophistication that made Amy feel awkward in comparison. Nevertheless, Amy *was* smitten.

For an experienced man like Hans whose taste in the opposite sex usually ran to someone with a little more experience than a teenage schoolgirl with a 'crush,' the attraction took some time to develop, but whatever Evelyn's intentions were in initiating the 'innocent' introduction, she couldn't have foreseen the earth-shattering impact that it would have on her niece, for in Hans Arregger, Amy had met the great love of her life.

2

AMY JOHNSON: BACHELOR OF ARTS

Amy prepared for her first term at Sheffield University with her head full of thoughts about Hans Arregger. He had quickly become the dominant factor in her life, and a welcome distraction from the tedium of her last year at the Boulevard, which had seemed like an eternity to Amy, helping to confirm in Hans' eyes her continued status as a schoolgirl. She invested all her heart and energy into expressing her love for him, often with an intensity that he found difficult to match, and occasionally even a little embarrassing. The idea of freedom was another facet of university life that Amy was looking forward to; away from the constraints and the strict discipline of home she would be able to please herself.

Will Johnson was becoming increasingly uneasy about his daughter's rapidly developing relationship with a foreigner. There was still very strong resentment harboured against Germany because of the war, and although Hans was Swiss, his name sounded sufficiently German for these thoughts to take root. Even the Royal Family were not immune from the anti-Teutonic feelings which were rife at that time. Royal courtiers had advised King George V to drop the family name of Saxe-

Coburg and Gotha, and so the House of Windsor was created.

For 'Ciss' there was the simple fact that Hans was Roman Catholic, and in her eyes, this above all else rendered him 'unsuitable' for any daughter of hers.

Amy's first task at Sheffield in the autumn of 1922 was to sit an entrance exam and to her dismay, she failed in every subject that she had chosen to study. The thought of her university career ending so swiftly, and in such abject failure, horrified her. Her feelings were amplified by the prospect of returning home to Hull in some disgrace, and consequently suffer all the hurtful jibes that were likely to come her way, and also by the fact that she would have to take any menial job she could get to pay back the portion of the grant that she had already spent. The intervention of a sympathetic Dean saved the day; he outlined the options available to Amy and she opted to study for a degree in economics. The intervention of a sympathetic Dean saved the day outlining the options that were now open to Amy. She opted to study for a degree in economics. Amy had allowed herself to be carried along with the idea of following Lizzie De Velling's example, but teaching had never really inspired her. It was the idea of going to university which had appealed to her most, so when the Dean offered her a reprieve, she grabbed it enthusiastically.

The halls of residence were already full, so Amy was forced to find lodgings with an 'approved' landlady, sufficiently close to the university, who kept a watchful eye on the students, most of whom were living away from home for the first time in their lives.

Amy had resigned herself to keeping in touch with Hans by writing him long, newsy letters, often running to several pages. She even wrote to him in French, but abandoned the idea when he sharply criticised her competence in the language: Amy reverted back to English, and reminded him somewhat pointedly that his English was heavily peppered with 'slang' words and could do with some improvement.

These letters gave her the opportunity to reassure him of her feelings, although she was often uncertain about the strength of his feelings for her, and her letters implored him to make some declaration or grand gesture in this regard. The reality of their 'forced' separation with him working in Hull and her studying in Sheffield only served to increase her sense of anxiety.

Amy had always been emotionally highly strung and given to drastic mood swings, with the tendency to wear her heart on her sleeve. Sometimes it was Irene who bore the brunt of Amy's frustrations in a fit of jealous rage. Even though the sisters had always been close, Amy was glad whenever Irene came to visit her digs at Sheffield. She had admitted to feeling homesick, but tried to remain cheerful and upbeat. She loved dancing to the gramophone and as the male students heavily outnumbered the females she was never short of potential dance partners. Even so, she told Hans in one of her letters that although the boys were always so generous with their compliments, they weren't entirely to be trusted, and that she would rather have had just the one dance with *him* at the Varsity Ball than a lot with those specimens.

Amy discovered in the first weeks of term that she had to sit an exam shortly before the Christmas recess. She was dreading it, especially after the humiliation of her entrance exam. She returned to Sheffield in January 1923 to collect her results, and was horrified, that she had failed again. Feeling embarrassed and wretched that she had let her father down, she went to plead her case with the Dean, but the faculty wouldn't bend the rules for her.

Her attendance at lectures had only been sporadic, and these results would affect the kind of degree she would be able to take. Her letters to Hans continued steadily throughout the early months of 1923. She was looking forward to the summer holidays when they could spend more time together. She liked riding pillion on Hans' motor cycle, wrapping her arms around his waist and enjoying the sensation of speeding along winding country lanes with the wind in her hair. Although Hans was extremely busy with his business interests, by the time the summer recess arrived Amy had a holiday with Irene to look forward to. They went to the Isle of Man where Irene proved to be very popular with all the single young men they met.

Although she was still very committed to her beloved Hans, Amy was nonetheless jealous of all the attention Irene was receiving, particularly when one of Irene's new beaux interrupted the holiday by inviting Irene to Dublin for an illicit weekend.

Amy forged the most important and enduring friendship that she ever had with a girl in her second year at university. She met Winifred Irving after a lecture and they walked back to their 'digs' together. Amy was quickly impressed by Winifred. She was confident, extroverted and they enjoyed many common interests.

One of Winifred's main interests was her involvement with the National Union of Students. She was an avid debater, attending inter-university debating sessions across the country and she urged Amy to take part. Gwyneth Roulston who had attended the Boulevard with Amy and Constance Tupholme, or 'Tuppy' as she was affectionately known, completed their quartet. They spent hours dancing, listening to the gramophone or just talking, and went on frequent camping trips to the nearby Peak District at weekends and during the holidays.

Amy's friendship with Winifred grew stronger with the years. They became soulmates as Amy developed the kind of bond with Winifred which up to that point in her life she had only ever enjoyed with Irene. Winifred was one of the few people to stay with Amy at the Aerodrome Hotel the night before she set off for Australia.

When Amy's name was later splashed across the headlines and Professor Knoop was casually asked if he remembered

Amy Johnson from her days at Sheffield, he replied mildly 'Was she here?' The lecturer responded by saying, 'Yes, she was Winifred Irving's shadow...'

Just as the shy Amy Johnson had once been content to bask in the reflective glory of the vivacious Winifred Irving during their university days, so Winifred would later accept the inevitable reversal of their respective roles, and take great pride in extolling the flying achievements of her greatest friend.

By 1924 the novelty of university had lost some of its appeal for Amy. Whilst her friendship with Winifred Irving and Gwyneth Roulston had helped her assuage the feelings of homesickness, and loneliness that had beset her during the early days at Sheffield, as well as the upheaval of moving from one 'digs' to another, there were still the nagging doubts she harboured about the strength of Hans' commitment to her.

The prospect of losing him horrified her. Hans' natural 'coyness' frustrated and often infuriated her when she compared it to her own full-bloodied commitment to the relationship and to the future that she had dreamt they would share. In the 1920s, a girl's future was enshrined in the institution of marriage, and the importance of a 'good match' could not be overstated.

The Great War had decimated the supply of eligible young men and Amy was determined to hold on to hers. Hans' Catholicism, which she knew was of paramount concern to her mother, was not a sufficient deterrent to a 'headstrong' independent girl like Amy.

Although Hans Arregger was in no particular rush to make any formal commitment to Amy, he continued to exert his influence over her in other ways, and Amy was readily willing to heed his advice and guidance. As an adolescent she had devoured the adventures of Angela Brazil, but Hans encouraged her to read more mature literature, urging her to try the works of George Bernard Shaw and the gritty tales of working-class life in the 'Potteries' penned by Arnold Bennett. She later accepted

Hans' suggestion that she might also enjoy the sophisticated dramatic works of Henrik Ibsen.

Hans was also cajoling Amy into agreeing to take their relationship onto a more physical level. For Amy to sleep with Hans was a very big step, but one that she felt ready for. It also prompted her to believe that Hans was ready to make the leap to a formal declaration of his intentions, and that the 'gauche,' rather clumsy attempts she had made to spark some feelings of jealousy within him by intimating in her letters a talent for flirting might have achieved the desired effect.

Hans Arregger had demonstrated considerable skill in subtly sidestepping Amy's frequent invitations for him to visit her in Sheffield. Pressure of work had always proved a valid if somewhat lame excuse. She was always eager to show him off and meet her friends. A Varsity Carnival Ball to be held early in 1925 seemed a suitable occasion, and she implored him to come, without holding out too much expectation, that he would accept. To her surprise he did, and Amy was thrust into finding him a place to stay sufficiently close to her own 'digs' to satisfy any clandestine ambitions they may have had.

In one of her letters to Hans, in February, 1925 she was so excited about the impending Carnival Ball, telling him that she wanted to wear her best red dress that she wanted *him* there with her on his arm. In short, she wanted heaven!

The social side of varsity life was still taking up a lot of Amy's time and her studies continued to suffer. Her final examinations loomed menacingly on the horizon, along with the black cloud of failure, which simply wasn't an option this time. Amy had struggled with her studies throughout her time at Sheffield and failure in two examinations had pricked her conscience into attending more of her lectures, but she still tended to cram her revision into marathon weekend sessions and then await the results in a heightened state of anxiety.

In May Amy learned that Hans was planning a holiday

in Switzerland,and dared to hope that he would mention his English girlfriend to his mother and sisters. She also moved 'digs' again to a small Derbyshire village called Hathersage which was to offer her the quiet and solitude she needed to pass her finals and get her degree. Amy was in a perpetual state of anxiety in the weeks leading up to her finals. Her fragile state of mind was not helped by Hans' absence while he was in Switzerland visiting his family and catching up with old friends, some of whom inevitably included former girlfriends. Hans had not, however, forgotten her completely – the frequency of her letters saw to that. He also advised her to try *Kola Pastillien,* a tonic he brought back from Switzerland for her which was designed to ward off feelings of lethargy as Amy crammed at every subject, right up to the 'eleventh hour' before the examination and then repeated the process with the next subject. *"What if I fail my degree?"* she wailed to Hans in one of her letters.

For all her anxiety and worry, however much a by-product of her own tardiness in neglecting her studies to pursue other interests, Amy proved in the weeks leading up to her finals that when the odds were stacked against her she found the reserves of strength and force of will necessary to see her through. It was to become one of the character traits that would win her *so* many admirers in the years to come.

Then one Saturday morning early in July 1925 despite a last minute scramble and with a borrowed £5 note Amy collected her degree. She was now a Bachelor of Arts in Economics and was justly proud of what she had achieved.

Although the degree would become something of a poisoned chalice setting her apart from other less well-educated girls, with whom she found herself working in the office environment, Amy returned home to Hull, a graduate and, having just turned twenty-two, allowed herself to bask momentarily in her own glory and to enjoy a long hot summer with the man she loved.

Will had business interests to pursue in America in the summer that Amy graduated from Sheffield, while 'Ciss' was taking Mollie and Betty to Bridlington for the time he was away; thus the family home in Park Avenue was shut, forcing Amy and Irene into lodgings on Anlaby Road. Before he set off on his Atlantic crossing, Will agreed to finance Amy's enrolment on a shorthand typing course. It was a useful skill with which to prepare herself for the world of work.

While Amy would have preferred to laze away the summer with Hans, she was acutely aware of the need to start earning a living. University had proved to be a very expensive experience for her and she was deeply shocked by the size of the debt that she had accrued during her three years of study.

Some of it had been the cost of board and lodging, although she had her grant towards the cost of essential living expenses. A large chunk had however been frittered away indulging her taste for expensive silk stockings and party frocks. Hans had agreed to loan her some money but this had not been some unconditional gesture of love on *his* part. He had wanted some kind of collateral, so Amy was forced to sign over her war bonds. After her brief course at Woods' Shorthand and Commercial College Amy responded to job ads in the local press, and she quickly discovered there was precious little around for someone who was in no position to be too particular or choosy.

She finally secured employment at Halls' Accountants on Bowlalley Lane, close to Hans' place of work and his flat in Wellington Chambers. Ever since her return from Sheffield, Amy had avoided her mother's pleas to join her and her two younger sisters in 'Brid', preferring to stay in Hull to be near Hans and listen as patiently as she could to Irene's often boastful 'plans' which ranged from a career on the stage to running off to Paris with her latest lover. Amy knew Irene could be as independently-minded as she was, but was yet to be convinced that any of her sister's plans, for a glamorous life amounted to anything more than wild fantasy. She didn't like the strict rigidity of office hours and was ostracised by the other girls who believed that Amy considered herself better than them with her degree and her sophisticated European boyfriend.

Amy also discovered that in spite of her crash course at Woods' her shorthand fell short of the standard expected at Halls', which was being met by the other girls. She often had to go over her notes again and again to make sure she had got it all down correctly and sometimes she felt the sting of humiliation and acute embarrassment when having to ask for certain words to be filled in for her before she could type her reports.

Matters came to a head towards the end of September 1925 when Amy collapsed at work, and had to be sent home. Will, who had returned from his three-month business trip in America, and 'Ciss' both feared that Amy was going to suffer a nervous breakdown like Irene had, while Amy's sudden loss of weight had convinced 'Ciss' that their eldest daughter was deliberately starving herself to look slimmer.

'Ciss' had decided that Amy should eat to regain her strength with good home-cooked fare and then be sent somewhere to convalesce. Bournemouth was mentioned and Amy immediately imagined herself ensconced in some opulent seafront hotel in the fashionable Dorset town, where she could recuperate in some style and Hans could arrange to visit her. Her hopes however were soon dashed as arrangements had been made for Amy to be 'billeted' with the Eddisons who were distant cousins of Will's and lived in a large comfortable house about a mile from the town.

In Bunty Eddison, a precocious five-year-old who latched onto her from the moment she arrived in Bournemouth, Amy experienced her first dose of heroine-worship. She was subjected to the child's constant chatter from the very start and Bunty followed her everywhere.

The Eddisons were house-hunting and gently insisted that Amy should go with them. In every other respect her time was her own, the priority being to recover.

She had her breakfast on a tray in her room and she sometimes didn't go downstairs until lunchtime. In some respects Bunty's admiration was just the fillip that Amy needed as her confidence and self-esteem were bruised by the cruel bitchiness she had suffered at Halls' Accountants.

Amy spent just over a month in Bournemouth and the gentler autumn climate meant that she could go for long walks occasionally, blissfully alone with her thoughts but often with the ever-devoted Bunty in tow. One morning they strolled together in a near-deserted Meyrick Park. Five years on Amy would visit the same park again as part of her nationwide tour; and a large crowd flocked to see and cheer the typist who had flown to Australia. For all the adulation that Amy would receive from an adoring public in the years to come, none could compare to the unselfish devotion of Bunty Eddison who showered on the fragile Amy a simple human kindness blessed with childlike innocence and exuberance which in its own unique way was the best tonic that she could ever have wished for.

3

HANS ARREGGER: HER ERRANT KNIGHT

Amy returned to Hull from her convalescence at Bournemouth in November 1925 to find the family home on Park Avenue a hive of activity, buzzing with excitement in preparation for Irene's twenty-first birthday. Amy was jealous as plans for Irene's coming of age were far more elaborate than what had been planned for her own passage into adulthood the year before. Amy felt her rightful position within the family as the eldest daughter was being usurped by her sister.

Irene appeared to have calmed down and the grandiose plans about living a bohemian existence and running away to Paris, had petered out. Whether this was due to her burgeoning relationship with Edward Pocock is unclear, but Teddy, as he was invariably known, had a steadying influence on the often skittish Irene, for which the frequently exasperated Will and 'Ciss' felt considerably grateful. The fact that he was the son of old family friends and made to feel welcome within the Johnson home to an extent that Amy's beloved Hans never would be, provoked Amy to frustration and anger, as she saw it as yet another sign that, despite all the anxiety and worry that Irene's volatile behaviour caused within the Johnson home, she still won favour with their parents.

Amy's main priority was to secure employment that would bring her enough money to meet her commitments at home, and relieve some of her burgeoning debt, the worry of which had kept her awake at night, and still leave her with some money over for fun.

Despite the occasional spate of jealousy, the affection that Amy had for her nearest sibling, towards whom she had always felt so protective, remained strong. Amy fretted about not being able to buy Irene a suitable birthday present befitting the occasion, due to the perilous state of her finances. As Irene devoted more of her time to her fiancé, Teddy Pocock, Amy sought the companionship of her younger sister Mollie – to whom she was also very close, despite a nine-year gap – especially as Hans' business interests were taking him away from Hull more frequently, sometimes to London. His trips afforded Amy the excitement of a clandestine weekend away, but invariably took him further away to the Continent, forcing Amy to revert to her old habit of writing him long, newsy letters to assuage her sense of loneliness.

Amy found a niche for herself in advertising when she joined Morison's Advertising Agency on Albion Street as a secretary on thirty shillings (£1.50) a week.

'Ciss' Johnson had made it clear in letters she had written to Amy during her convalescence at Bournemouth that she would have to find a job on her return to Hull as she would be in the way at home and could easily become bored. In *Myself When Young* Amy describes how she would study advertisements in the local press, where she had found the job at Morison's advertised, making copious notes on how the layout could be improved.

Morison's was the largest advertising agency in Hull, taking their professional ethos from the large American agencies on New York's Madison Avenue rather than their counterparts in London. Hans had carefully suggested that Amy take the opportunity to work in America as a cooling-off period but

Amy, buoyed up by the endorsement that Hans had received from the Eddisons, was having none of it. While Will and 'Ciss' hoped that any time apart would end the infatuation between Amy and her Swiss boyfriend, Hans was still ambivalent about the true extent of his feelings for Amy. While her unqualified devotion obviously flattered his ego, her emotional, clinging instinct stifled him and he felt trapped by the sheer force of her passion. Amy still harboured strong hopes of marriage, telling him in her letters of how blissfully happy they could be.

She had visited the Ideal Home Exhibition on a trip to London and was ever keen to learn the finer points of housewifery, and become the best wife that Hans could ever wish for.

At Easter 1926 Hans Arregger took Amy to Loch Lomond for a short break. Although she was delighted with her 'honeymoon' in Scotland, Amy still pestered him to take her to meet his family in Switzerland seeing it as a sign that he was claiming her as his girl by introducing her to his mother.

In May the country was in the grip of the General Strike. A nationwide walkout originally in support of the miners, paralysed the country while the first-ever Labour Government under Ramsay MacDonald, elected two years before, seemed hopelessly out of its depth, and appeared powerless to act. Hans was feeling worn down, but he finally acceded to Amy's demands, and in early summer 1926 he finally took her to his beloved homeland.

Amy was determined to make the best possible impression on Hans' family and she set upon this task from the moment that she and Hans arrived in Lucerne.

Marriage to Hans had become an obsession for Amy, especially given the speed in which Irene had elicited a firm proposal of marriage from Teddy Pocock after so short a courtship, whereas she was no further down the road towards matrimony after four years of dating her beloved Hans. This fact, which so irked

Amy, gave her mother reason to hope that the infatuation would eventually fizzle out of its own accord without Amy getting too hurt and Will and herself having to lay down the law and risk alienating their stubborn eldest daughter even further.

In this respect 'Ciss's hopes were to be seriously dashed as she had badly underestimated the strength of feeling Amy had for Hans and the lengths that she was prepared to go to keep him. Amy had managed to cajole Hans into agreeing that their future would be seriously discussed before her twenty-third birthday on 1st July, which she would be celebrating with him and his family in Switzerland. It was with this promise safely secured and with a lightness of heart and with her man on her arm that Amy arrived at Lucerne in June 1926.

Hans' mother was a tall and commanding figure in her widow's robes and her mass of steely grey hair. She appeared slightly formidable with her stiff demeanour and was very different from 'Ciss' Johnson, yet she took to Amy quickly although Amy tried very hard to please as if she were taking a test on which her very future with the man that she adored would depend. Hans' sisters were also at home, one of whom was on holiday from Paris where she worked as a governess and they too readily accepted Amy as their brother's future bride. The Arreggers were as piously devout in their Roman Catholic beliefs as 'Ciss' Johnson was to her strict Methodist ethos. Amy happily attended Mass on Sunday morning with the rest of Hans' family, an act that would have enraged 'Ciss', who would have seen it as a deviation that could not be easily forgiven. Amy was thrust into an impossible position, for between these two great pillars of religious adherence, each one as strong and as impenetrable as the other, she was just a girl of nearly twenty-three hopelessly and madly in love.

Amy enjoyed much of Hans' time and attention during her two-week stay in Switzerland. It was a rare treat for her to see so much of him during the day and although the presence of his family meant it wasn't quite as idyllic as her 'honeymoon' in Loch Lomond had been, she saw for herself the love and respect which Hans had for his mother and how seriously he took his role as the titular head of the family.

When Hans had appointments he had to keep, Amy enjoyed the company of his mother and sisters, who made every effort to make the English girl welcome in their home, believing that this was the girl that Hans might one day make his wife.

Frau Arregger made it clear to Amy, that she would be expected to convert to Roman Catholicism once she and Hans were married. Amy, buoyed up by the promise Hans had made her, believed this to be significant, and took the remarks regarding religion as a mother's blessing. Amy's lack of competence in speaking German piqued her somewhat as it meant she relied upon the Arreggers to speak English solely for her benefit. Hans introduced her to all his friends; and Amy learnt that one friend in particular, named Hermann had known Hans since childhood.

Amy celebrated her twenty-third birthday in Switzerland, and although it was unusual not to have her family around her to help celebrate the day being with Hans was for her far more precious especially as she anticipated that this would be the moment that he would choose to ask the question she'd been waiting for nearly four years. His friends like his family treated Amy as his future wife and not as just the latest in a long line of casual flings. This was reflected in the gifts that his friends brought for her birthday, which was celebrated with an evening meal at a restaurant in Lucerne. The wine flowed easily and Amy was happily content in the company of the man she adored. But the burning question had not yet been asked. She was forced to wonder whether Hans had deliberately forgotten the promise that he'd made to her before setting off on the holiday or whether he was simply waiting for a special intimate moment to ask her to become his wife.

Even as a young woman she still wanted the romantic fairy-tale prince she had read of so avidly as a girl and, for her Hans, had always fulfilled that role. As the evening wore on Amy's patience began to wane as Hans, heavily imbued with large quantities of alcohol and in the company of friends he had known for years, exaggerated his affected continental manners, and behaving in her view quite outrageously.

As a teetotaller Amy did not indulge in the wine, but looked on as Hans kissed various females, in their company seemingly oblivious to her and her feelings.

Amy, stung by embarrassment, and an acute sense of betrayal, was determined to confront him, before the night was over. She returned home bitterly disappointed that Hans had failed to keep his promise to her and cement their future together with a firm proposal of marriage.

In a private intimate rendezvous on the steps of his mother's apartment, in one of the rare moments they had alone together during their time in Switzerland, she had anticipated that this was the idyllic moment which Hans had chosen to do the honourable thing, but to no avail. They returned to Hull via Paris and a trip to the horse-racing at Longchamp in a tense and brooding atmosphere.

Amy felt that Hans was treating her ever fragile feelings with an attitude of mild disdain. So, feeling emotionally raw, she decided to vent her anger in the usual way by writing Hans a long letter, in which she laid bare the sense of betrayal and humiliation she felt adding that she wanted to smash the vase that his friends had bought for them as a token 'wedding' gift, as it had come to symbolise the lie on which their relationship now seemed to be based.

Her job at Morison's offered her scant consolation, but provided her with a distraction of sorts, as she believed that in advertising she had discovered a niche for her creative talents. Although she felt some residual resentment by some of the girls in the office at what they perceived to be privileges enjoyed by her, it was a mild torment to endure compared to the spite and bitchiness she had been subjected to in Frank Halls' Accounts office.

While several male members of the staff at Morison's valued Amy's contribution, her diligence and hard work, she still felt restless and frustrated at the monotonous nature of some of her duties. Morison's was the largest agency of its kind in Hull and the most prestigious occupying one of the grand Georgian houses on Albion Street. Amy believed that as she wanted to stick at advertising the prospects of climbing up the ranks would be better served in one of the large advertising agencies based in London.

Although she had only visited the capital a few times she was already referring to it as 'town' in her letters to Hans, whose own business interests were increasingly taking him away from Hull to Germany, Italy, and Switzerland. Amy steeled herself to writing to all the top advertising agencies in London, with the view of taking the drastic step of leaving her family and 'dreary' old Hull behind and sacrificing the security and comforts of home.

In the autumn of 1926 Amy found herself alone yet again, and seeking out her second sister Mollie for solace and companionship. A bond had developed between them, and it was with Mollie, one cold November Sunday that Amy had her first experience of flying. They had cycled out to a field in nearby Cottingham for a joyride over Hull in a five-seater Avro biplane: The joyflight was short and it cost them five shilling (25p). Mollie would have readily gone up again but didn't have enough money. Amy, however, was fairly unimpressed by this first experience; the so-called exhilaration of flying was lost on her, and she later remarked that the wind from the open cockpit had *'messed up my hair'*.

Life at home however was becoming increasingly intolerable for Amy. Relations with her mother were more fraught than they had ever been and Amy's continued 'infatuation' with Hans Arregger was inevitably the cause. Hans had perhaps, wisely decided to spend Christmas, 1926 with his family in Switzerland; and Amy wrote to tell him that although very disappointed that she wouldn't get to see him over the festive season, she would attend Mass on Christmas morning, as a way of feeling closer to him. In Amy's mind this natural act of communion with the man she adored was purely innocent, but to 'Ciss' it was a wilful act of betrayal and a severe test of parental authority by their eldest child, and encouraged by the Papist foreigner, whose very presence in Amy's life, was an anathema to the devoutly Methodist 'Ciss.'

When Amy was seen by a family friend leaving St Vincent's Roman Catholic Church on Christmas Day morning – the news finding its way back to her outraged parents – it set a dark gloom over Yuletide celebrations in the Park Avenue house, despite Will's efforts to the contrary for the sake of his younger daughters. Amy's fate was all but sealed when 'Ciss' urged Will to issue Amy with an ultimatum. Her family or her Swiss boyfriend. The long game of wait and see and hope they had so patiently played, simply hadn't worked. Amy still wanted to marry Hans Arregger. The fact that he'd dithered about proposing to her was a blessing of sorts, but this merely confirmed to 'Ciss' that his religious convictions aside, he was totally unsuitable for Amy, by leaving her dangling so cruelly for a marriage proposal that would never come; when she could be setting her sights on someone, with whom she had some genuine hope of a future, as her sister Irene had done with Teddy Pocock.

The fact that Irene seemed to spend so much time quarrelling with her fiancé, over 'trivials' such as curtains for their new home didn't encourage Amy that she could do much better. For a girl brought up on countless fairytales of 'fair-headed princes' rescuing maidens in distress; Hans was just the man. The fact that her parents still so strongly disapproved of him only made him for her, as attractive as ever.

Will Johnson had deep misgivings about the ultimatum idea. Knowing Amy's taste for rebellion in the past, he was very conscious that by forcing her to choose they could easily throw even her further into the arms of the German-speaking Swiss. The prospect of Amy banished from home, living virtually as Hans Arregger's mistress filled Will with a sense of horror. He was a local businessman of some standing with a reputation to maintain. This for him overrode all other considerations.

He chose to give Amy a choice that least appeared to offer her the chance of something better: a new life in Canada, where his brother Tom lived and would look after her.

Hans himself had once suggested that Amy go to work in America as a cooling off period when their relationship was at its most intense and her demands to him about marriage had been almost constant.

It was Mr Eddison, during her convalescence at Bournemouth, who having met, and been impressed by Hans, had urged her against it. Will hoped now in the early spring of 1927 that Amy would take this chance, sail off to Canada and exploit all the opportunities that it offered a young, well educated girl like her. Amy declined the offer. Astonishing her father with her act of defiance, which he considered be ill-mannered, ungrateful and extremely foolish.

Amy had been forewarned what this decision would mean: her exile and no further financial assistance. She was on her own. For Hans Arregger she had made the ultimate sacrifice. She had nailed her colours firmly to his mast, and pledged herself to a future with him that was still as yet unknown. She had also reached a critical point in her relationship with Will. An impasse. For them the father-daughter bond had always been very strong but at that moment in her father's study with a look of hurt, and deep, deep disappointment etched on his face, she knew that for both of them now there was no turning back.

4

EXILE IN LONDON

London in 1927 was an exciting place to be especially for a young woman from the provinces, whose only previous experiences of the capital or 'town' as Amy had so curiously described it, had been gained on short trips, to the Ideal Home Exhibition, or on the clandestine overnight stay in an opulent West End hotel, that she had enjoyed with Hans.

The decade-long party once seen as the perfect antidote to the misery and the carnage suffered in the trenches, had past its halfway mark, and was destined to fade into history as the '*Jazz Age.*' Yet the hectic pace at which some people lived, showed no sign of abating. It would take a long and deeply entrenched world Depression, before the collective hangover of so much champagne, and unbridled gaiety finally kicked in.

Amy arrived in the early spring slightly shocked by the enormity and the finality of the decision she had made, but also very excited. The bright lights of Piccadilly dazzled her; rending her wide-eyed, at the wonder all around her. She had always loved dancing to the gramophone, during the evenings at Sheffield and now she could dance the night away at the Trocadero to the sounds of the 'big' bands.

There was certainly no shortage of erstwhile young men eagerly jostling to escort her on the dance floor and around the decadent 'hot spots' of a tumultuous city, whose people were buzzing with the sheer joy of just being alive. Her main concerns however were to find gainful employment, and a permanent place to stay. The YWCA hostel in Bloomsbury provided a roof over her head in the short term, but the strict atmosphere in the long, drab dormitory rooms with little privacy, was depressing. The lights-out policy at 10.30 pm was rigorously enforced which meant that Amy was forced to write her long letters to Hans by torchlight under the bedcovers.

Amy's early efforts to find work had not proved fruitful, and with her reserves of cash dwindling, she grew more frustrated. She had written to several top London advertising agencies before leaving Hull, without much success. The ones who had bothered to reply did so with a curt rejection letter which plunged her fading spirits further. She was also very conscious that the little cash that she had couldn't just be frittered away on luxuries until she was earning again. As part of his ultimatum, Will Johnson had warned Amy that there would be no more financial help; yet he had with an air of sad resignation and a sense of profound helplessness given her £10.

Her burning ambition to leave the YWCA hostel and find 'digs' was seriously hampered by the alarming lack of employment opportunities coming her way. Forced to accept the reality of her drab living conditions a little longer, Amy welcomed her aunt Evelyn Petrie to London. Evelyn who had introduced Amy to Hans Arregger, came down to check on Amy's well-being and bring her news of home. Evelyn told her that her Johnson grandparents were deeply dismayed by Will's decision to adopt such a hard line with her, adding that they didn't approve of his methods. It was good for Amy to have this link with home so early in her London adventure, especially as it was so unexpected. Evelyn Petrie left Amy in much brighter spirits than she had found her.

Amy had her share of male admirers, eager as ever to compliment her, and shower her with small tokens of their affection. They clearly had an ulterior motive, besides which her heart still *belonged* to Hans Arregger, perhaps more so than ever, as she had now endured exile rather than give him up.

She was however dismayed and disappointed by his reaction to her decision, and was forced to wonder whether he was somewhat relieved to have her out of his way in London.

Amy was to annoy and hurt her bewildered parents still further when she made the very difficult decision not to attend Irene's wedding to Teddy Pocock scheduled for late April, 1927. The ceremony was later postponed until the end of May when Irene was struck down with a serious bout of scarlet fever weeks before her nuptials and had to be admitted to a sanatorium to recover. Amy hated the thought of deserting the sibling with whom she had once been so close. In her memoirs *Myself When Young* Amy later described how she had been fiercely protective of her younger sibling who had looked so fragile as a child.

Amy's real motive for staying away was totally unselfish. She felt that given the atmosphere at home when she had so abruptly left; to begin her 'exile' had been so intense and bitter; that her presence at the wedding could only spoil Irene's big day.

It was on the third Monday in March, that Amy having finally secured employment began work at Peter Jones on the company's Learnership Programme in the Silks Dept, of their impressive store on the King's Road, close to Sloane Square. John Lewis had bought the store in a cash deal and handed it over to his son to 'do something with'.

It possessed an upmarket feel, attracting the most fashion-conscious women through its doors. Amy had never seriously considered a career in retail sales as advertising was where she had wanted to be. Once again her impetuosity had let her down. She had 'committed' the *ultimate* sin of not studying the small print of her Learnership contract. It was an elementary omission which would have made an astute businessman like her father, cringe with embarrassment at her naiveté. When Amy received her wages the end of her first week she quickly discovered to her horror the enormity of what she had taken on. The Learnership Programme levied a charge for the services to the company, based on the employee's value to them in that week. Amy's services were valued at just ten shillings (50p) and so in one week Amy had accrued a debt of £2 and ten shillings.

Amy stuck with the Learnership Programme for just over a month, which was about as much as she could endure. It was an exhausting and dispiriting experience, her sense of disillusionment, at its most acute at the end of each week, when she received her wages, and discovered how much her services had been valued at for that week *and* how much more debt she had accrued to Peter Jones as a result.

She wrote to Hans, saying that her job as a salesgirl required virtually no brain power, but she was left physically exhausted at the end of each day by the constant running about. As a new girl Amy was inevitably given the mundane errands to run that were considered beneath the senior staff. Olive Birbeck, a step cousin of Amy's from Leeds lived in London and worked at *The Times Book Club* on Wigmore Street made contact with Amy to check on her well-being in much the same way that Evelyn Petrie had done.

She secured an appointment for Amy with her employer, but Amy, desperate to leave Peter Jones, was disappointed when it led nowhere and she considered it to have been a waste of time. Vernon Wood, an old family friend, invited Amy along with Olive Birbeck to dinner at a very refined and expensive restaurant in Holborn. He ran his expert legal eye over her Learnership contract with Peter Jones and concluded that it was hopelessly one-sided in the employer's favour and highly unsatisfactory. As a senior partner in the legal practice of Messrs William Charles Crocker, Wood was in a position to offer Amy a job as a typist, adding that there was scope for her to perhaps become *his* personal secretary, once she had gained the necessary legal experience. Amy, eager to realise any means of escape from the wretched Learnership Programme that she had come to despise, jumped at the chance that Vernon Wood was offering her. Their grand, imposing offices at 21 Bucklersbury, were close to Mansion House in the heart of the city and so Amy's chequered career took *yet* another turn as she entered the dusty world of the law.

Having helped Amy to escape the drudgery of the Silks Dept at Peter Jones, Olive Birbeck had now found a possible solution to her spartan living conditions. She knew of a middle-aged couple living in West Hampstead, who were willing to take in a lodger, as long as they could find someone suitable to invite into their home. Amy's long-term ambition was still to find a London home she could share with Hans, but that prospect was looking less likely than ever and in the immediate future, a viable escape from her drab surroundings at the Bloomsbury hostel was what she craved most of all.

Amy had been at Crockers just over a month in May 1927 when Charles Lindbergh made his historic solo crossing of the Atlantic, in the easier west-to-east direction. Lindbergh was to earn the distinction of being the first *true* flying 'superstar.' He was the pioneer who blazed the trail in solo record-breaking flights across the Atlantic and the first to make serious money from his flying exploits thanks largely to an over-blown media circus which coveted pioneer aviators as though they were film stars to whom headline-seeking press barons on both sides of the Atlantic were willing to pay big money for the privilege of printing their story.

Amy's main concern was adapting to the demands of the

law. She took to her new environment very quickly. Vernon Wood was very impressed by her attitude and soon he was giving her extra duties and more responsibility which Amy relished, even though it invariably kept her at the office well past 7 pm.

By September that year when she had been with the Crockers just over five months, she became Vernon Wood's personal secretary. She wrote to Hans telling him the news of her promotion, and the rise in salary that came with her new role. She was adapting to life in the metropolis much better than she had imagined and no longer felt as intimidated by the enormity of her surroundings. *Yet* she longed for Hans to be here with her. On a business trip to Lisbon he came via London to see her.

The last three months of 1927 was the most life changing tumultuous and in some respects the most daunting period she had experienced. They passed by in a flurry of frenzied activity. She was on relatively good terms with her parents once again and although the thaw in relations was not as yet complete both Will and 'Ciss' had been to London to see her, and Amy had made it home to Hull for her parents' silver wedding celebrations in September. While there was a tacit, unspoken acknowledgement that Hans Arregger was still on the scene, the reparation of the Johnson family unit had begun.

There had been tentative plans laid for Amy to spend Christmas 1927 in Switzerland with Hans and his family, but by the time came the plans had fizzled out to nothing, as had their proposed skiing trip to St Moritz early in the New Year. Hans went on without Amy with pressing business matters weighing heavily on his mind, and causing him much anxiety.

Amy and Winifred Irving had concerns of their own to occupy their minds. They were forced to move out of their small flat on Marylebone High Street, due to constrained financial circumstances on Winifred's part and move into Ames House;

a drab, colourless place on Mortimer Street close to Oxford Circus. Another YWCA hostel was something of a step back for Amy, having just experienced the comforts and the ambience of home as a lodger in West Hampstead, which her step cousin, Olive had found her.

Winifred Irving had come back into Amy's life when her job at the Women's Guild of Empire came to an end, and she secured a new position with the biscuit makers McVitie's in the Welfare Dept of their Willesden factory. The bond between them was as strong as ever and in a bid to recreate the carefree, joyful atmosphere of their days as students, at Western Bank Sheffield, Amy abandoned her cosy West Hampstead billet and they took a small flat in Oxford and Cambridge Mansions. Amy and Winifred endured the awfulness of their surroundings at Ames House for as long as they could. As 1927 faded out giving way to 1928, they preferred staying out in the evenings after work visiting their favourite nightspots until they ran the very real risk of being locked out for the night when they returned to the depressing dormitory rooms which resembled some cold, remote girls' boarding school. By the end of February, however, they had found themselves a new home at 24 Castalain Road, Maida Vale. Soon after settling in, Amy took off on holiday. Walking alone along the sand dunes at Perranporth in Cornwall, and often deeply reflective in mood, she allowed herself the time to reassess her relationship – if it could still be described as such – with Hans. It pained her considerably given the time and effort she had invested into Hans right from the very beginning, that to have received so little from him in return left her so totally bereft. She felt like one of the many empty seashells that she found on the beach and retrieved for her youngest sister, Betty.

Amy had the palpable sense that Hans was moving irretrievably away from her, and that it would take everything, that she had to keep him now.

She knew that in the past a pleading letter from her, in which she laid all the difficulties their relationship experienced at her feet, would be sufficient to get them back for the time being at least on an even keel. It is significant, judging by the letter that Amy wrote to Hans on her return from her holiday in Cornwall, that the week away had changed her, and the perspective from which she now viewed her relationship with the man she had once loved so completely.

Hans had made Amy aware that he had acquired a new friend, who he was escorting to the theatre, cinema and to dances. Amy received the news with as much good grace as she could manage, referring to her in one letter as your 'new sweetheart'. She chose to remind him of every conceivable sacrifice that she had made so that they could be together while Hans, for his part, had over a period of seven years yielded precious little to her in return. She added that he was not to blame for the fact that she had put him on a pedestal and that he had attained a god-like status in her eyes, which not only served to flatter his ego, but that it had made him view her as some kind of obedient slave always willing to bend to his will, eager to do his bidding.

The most devastating line in that letter was Amy telling him Hans that she no longer wanted him sexually. Hans had initiated the physical aspect of their relationship and Amy had acceded to his demands for sex. As for her the desire for sexual intimacy was a symbol of love, and because she no longer believed that they could be happy together.

Whether this outpouring of pent-up emotion was a sign of acceptance on Amy's part that after seven long years the game was finally up with Hans and that they should go their separate ways *is* impossible to say. Her brave words could have been written in the hope that he would repudiate them and that in her next letter she would slavishly express her gratitude and, that all was well between them again – at least for the time being. Amy did tell Hans that she had been a fool; forever dreaming of the fairy-tale 'happy ending', waiting on him for soothing words in letters but now he wrote less frequently.

Amy was not, however, even remotely prepared for the possibility of Hans marrying someone else. One hot sultry evening in mid-July 1928, shortly after her twenty-fifth birthday Hans called on Amy at Castalain Road, and invited her out for the evening. She was delighted to accept his invitation, albeit surprised even though they had been out just as friends since she'd written the fateful letter and he'd acquired his new sweetheart as Amy chose to call her.

Hans had invited Amy out to give her the devastating news that on 14th July 1928 he had married Connie Richards. His journalist friend from Liverpool was now his wife. Amy cut the evening short and returned home, shocked and deeply hurt by the news. She threw herself onto her bed, sobbing uncontrollably through a torrent of tears. This was truly the end game. Whatever dénouement Amy had envisaged for herself and her often troubled relationship with Hans, even in its darkest moments never had she imagined this.

Amy only ever met Hans Arregger one more time after that fateful July evening when he told her of his recent marriage. It was after her Australia flight when she was the woman of the moment, feted wherever she went by an adoring public. She was in Hyde Park and there was the handsome Swiss; his wife; and the ageing Frau Arregger, his mother who was on holiday from Switzerland. The small group chatted happily for a while, and then departed.

Amy never mentioned Hans by name in her memoirs *Myself When Young* referring to him only as a boy, that she had become infatuated with, but that it hadn't worked out, adding somewhat obliquely that her parents, who'd had their doubts from the start, knew her better than she knew herself.

In 1982 the Arregger family put the letters that Amy had

written to Hans, over 300 in all covering a period of seven years, up for public auction.

Hull Central Library snapped them up, and they remain their intellectual property, representing a small slice in the personal history of the city's most famous daughter.

5

EARNING HER WINGS

In mid-June 1928 the androgynous tousled-haired, free-spirited feminist Amelia Earhart became the first woman to fly the Atlantic, albeit as a passenger and casual observer — nobody was more dismissive of her contribution to the feat than Earhart herself describing her presence on board as being as much use to the pilots as a sack of potatoes and yet when the *Friendship* touched down at Burry Port, Wales, crowds flocked to catch a glimpse of the icon that was Amelia Earhart.

The press were charmed by her unassuming personality as was the American-born female MP, Nancy Astor, who despite being somewhat unimpressed by Amelia's deliberately unfeminine appearance, escorted her around London where she gave speeches to several pressure groups, all of whom expressed their approval of the social worker-cum-pilot who was doing so much for women's causes on the other side of the Atlantic. However these factors apart her very presence on the flight, demonstrated one sober point all too graphically; the extent to which Britain was falling behind her main competitors in the race for dominance of the skies.

For Amy, flying was the new fascination in her life and one

she came across quite by accident. Boredom and a sense of lethargy had prompted her to board a bus bound for Stag Lane aerodrome one Saturday afternoon in late April 1928. She had been eager to join some kind of club for a while and a tennis club seemed an obvious choice.

The tennis courts were visible from her bedroom window, but the overall setup was poor. The spartan amenities attached to the club failed to inspire her and in Amy's view a social club which couldn't run to the occasional dance was a very poor show.

The cost of flying lessons was exorbitant for a solicitor's secretary and the waiting time stood at six months, yet Amy, wide-eyed with wonder, was not going to be deterred. She took to the atmosphere at Stag Lane like a moth to a flame. For the fish merchant's daughter from Hull, the years of mindless drifting were over, and in the years to come Stag Lane became almost like a second home, as at last she had found her metier, and the era of the Lone Girl Flyer had begun.

Amy had her first flying lesson in mid-September 1928 and in spite of how much she had been looking forward to it, and the length of time that it had taken her to get airborne, it was not a happy experience. Her instructor, Captain Matthews was not impressed. As Amy had trouble in hearing, and understanding the instructions that he was barking at her down a tube from the front cockpit, she felt an acute sense of frustration. She was no casual weekend flyer. The lessons were expensive, but Amy was eager to learn the basics as quickly as possible. Overall her tuition was sporadic with long gaps between lessons, while she saved up more money.

She had cut her living expenses to the bone to afford her first lesson, and further cuts followed in due course. She was determined to prove that she was not just another dreamy *flapper* for whom flying was the latest fad. It had rapidly become her sole obsession, and for her no sacrifice was considered too great, in the bid to become a fully licensed commercial pilot.

Captain Matthews was not inclined to overt praise and encouragement to those he was teaching to fly, so as the days of autumn slid gently towards winter, and the prospect of fog rendered flying difficult and often impossible, Amy's frustrations grew stronger. Her determination and enthusiasm had never been in doubt but were not matched by the means to fund regular lessons.

By November she was taken up by a different instructor, with whom she felt more at ease. James Valentine Baker was a veteran of Gallipolli and the holder of a Military Cross. He relied for communication with his pupils on carefully rehearsed hand signals rather than the dreaded speaking tube, and he encouraged Amy to practice her turns although he was also inclined to think that she wasn't yet ready to go solo.

Her father Will, on one of his business trips to London, had been relatively encouraging about her plans to learn flying. Amy planned to spend Christmas at home in Hull. She had long since mended the rift with her parents as the final split from Hans, and *his* subsequent marriage to another woman, had only helped to convince 'Ciss' that her original instincts regarding the Papist Swiss had been right all along. So Amy's rightful position as 'heiress-apparent' was happily re-established.

Amy enjoyed her family Christmas at home in Hull, and returned to London, to welcome in the New Year of 1929 relaxed, and imbued with a new sense of vigour, determined to master the finer arts of flying, which had so far eluded her. Circumstances conspired against her. Adverse weather conditions kept her grounded throughout most of January, and her sense of frustration grew as sickness, and changes in personnel, meant her flying hours in the first quarter of the New Year were as sporadic as they had been in the last quarter of 1928. When the worst of winter eased sufficiently to allow her to go up it was with James Valentine Baker in the front cockpit. Baker, whose teaching methods she preferred, was biding his time at Stag Lane, having already secured a post at Heston aerodrome. So Amy was again at the mercy of the fastidious Captain Matthews until the new chief instructor, Major Herbert Travers, arrived at Stag Lane.

Major Herbert Travers struck an imposing figure. At thirty-eight he looked older than his years, and he had suffered an horrific injury to his right arm in the trenches during the Great War. A stretcher party had taken an enormous risk to rescue him in the face of enemy fire, and in spite of his injuries he returned to France, and was later awarded a Distinguished Service Cross.

He attained the rank of Major in the newly formed RAF after the war and arrived at Stag Lane after a spell at the Cinque Ports Flying Club based at Lympne on the Kent coast, where he'd held a temporary position.

Among his many notable pupils was the publisher Hamish Hamilton and Francis Chichester who later found fame as an accomplished yachtsman. Aside from Amy, he also had many female pupils, one of whom was the diminutive New Zealander Jean Batten, the daughter of an Auckland dentist who had come to England to learn to fly along with her fiercely overly protective and ambitious mother. Largely because of her almost *obsessive* need for privacy combined with her stunning film star looks, she earned herself the sobriquet, the '*Garbo of the Skies*'.

By the time Amy was considered ready to go solo it was April, and with it came the first vestiges of spring. Then dense fog dogged her progress, together with her difficulty in achieving a 'textbook' landing. This was always the one aspect of flying that never came easy to Amy. The relationship between Amy and Travers was never an easy one. He found her petulant, with a point to prove. Amy felt acutely self-conscious in his presence, especially about her Yorkshire accent. She was back with Matthews in the early summer of 1929 and in June Amy went solo for the first time in a *Cirrus Moth* and a month short of her twenty-sixth birthday, she had finally taken the first tentative steps towards a much coveted career in aviation.

Apart from the thrill that flying gave her – a kind of elixir of which she simply couldn't get enough of – Amy also took to the social aspect of her membership of the London Flying Club, with an ease that surprised her. She felt that she could just be herself with fellow club members, and had nothing to prove. She was trudging off to the pavilion at Stag Lane most weekends with its promise of a roaring fire, and cups of steaming hot tea to thaw the chill in her bones. It was like an oasis in a desert of gloom as the long harsh weeks of winter edged slowly on, and the adverse weather conditions, which kept her grounded stretched her patience to the very limits of endurance.

Amy relished the clubhouse banter and was eager to pick up any useful tips and scraps of advice so she relied on the benefit of experience of those who had been flying for years. It was grist to the mill for an 'apprentice' flyer determined to learn everything she could to make her the best pilot she could be.

One of the London Flying Club members with whom Amy became particularly friendly with was Will Hay, the music hall star, who was to find fame in a series of uniquely British-made short 'slapstick' films in the 1930s and 40s, which at the time were the staple output of the fledging British film industry. He seeped into the collective conscience of a cinema-going public as a middle-aged man who seemed to be having the time of his life, acting the fool on screen. There was, however, a serious side to him, and although flying was largely a pleasurable hobby, for which he kept his Moth at Stag Lane, the theatre would always be his first love.

For Amy the relationship was purely platonic. Will Hay was forty-one and married, so negating any prospect of romance. The bond of friendship between them however was later deemed strong enough, for her to introduce Hay to her father. As time passed, Amy and Will Hay's ongoing friendship raised a few eyebrows, and set tongues wagging in the clubhouse, but for her it remained innocent and it was a long-held, genuine love of theatre which prompted her to go and see him at work on the London stage, and to accept his invitation to go backstage afterwards.

Although Hans' callous betrayal of her was fading somewhat into the past, her emotions were still raw, and she was yet to meet anyone who held out any serious prospect of winning from *her* the deeply held affection that the Swiss had once enjoyed. Having secured her 'A' licence, Amy's mind was now set on steering her towards a commercial 'B' licence and latterly towards securing her tickets or the 'C' licence as a qualified Ground Engineer. At that time only one other woman had ever achieved the latter. Lady Heath, having worked as a mechanic in the United States, had allowed her tickets to lapse in recent years, so by the time Amy passed the necessary examination late in 1929 she became the only woman in the world to hold a valid Ground Engineer's licence.

In just over fourteen months Amy had attained all three major licences, a fitting and satisfying rebuttal to her detractors who had scathingly dismissed the prospect of her ever making it as a pilot. The only obstacle to her securing a career in aviation now was her gender. Aviation was still largely a male dominion. It took a very exceptional woman with the determination and stamina to keep swimming against the tide. Amy was to prove she had all the 'pluck' required to get herself noticed.

Jack Humphreys, the Senior Engineer at Stag Lane, was to prove himself as a crucial ally to Amy in her bid to get her tickets and he became something of a gallant knight fighting her cause. Initially she had intrigued, and amused him. Amy was more than willing to exploit his position within the social structure of London Flying Club to get herself behind the hangar door and into his world, beyond which most male club members never ventured to go, let alone a woman.

Yet Jack liked Amy from the very start and he admired her grit. Soon she became a part of his team, donning overalls and grabbing a spanner.

She dropped her girlish first name and encouraged the men to call her 'Johnnie', an androgynous nickname that helped them to treat her like one of the boys.

The girl who had once relished boys' games at the Boulevard School would often kick a football around the tarmac at Stag Lane during the lunch hour and after work. The law and her job at Crockers, although still important to Amy as her only source of income, was however slowly becoming less so. It was staid and lacking in glamour compared to aviation. Flying gave her a thrill, an edge. Vernon Wood was still impressed with Amy as his personal assistant, and she still put in a fair day's work at the impressive suite of offices close to Mansion House. Her fingernails, once so immaculately manicured, now had oil stains under them, and she would often come into work out of breath from rushing to make it on time. She was still very much there in the physical sense, but her heart and her mind were elsewhere. It was the freedom of an open cockpit that stimulated now, with the wind to 'mess up' her hair, rather than the polished mahogany-panelled solicitor's office, and the comparatively safe, ordered world of the law.

Amy spent the summer of 1929 dividing her time between Stag Lane and her duties in the hangars which she loved, and her day job as a solicitor's secretary which largely through necessity, she had now merely come to *endure*. She longed to leave Crockers for a job in aviation, and had started outlining her intentions to leave that autumn after two and a half years. She hoped that by dropping subtle hints into the text of her letters, the news would seep slowly into her parents' conscience, but she knew that this would only be possible with the support, both morally and financially of her father.

Will Johnson had been immensely proud of what his

eldest daughter had achieved so far in aviation, but he was reluctant to commit himself to assisting her financially in what was still a precarious and fledging industry, full of potentially life-threatening dangers. Amy felt an acute sense of frustration at his procrastination, but realised that patience was the key to winning the day. Will had always been willing to help his offspring but he had always been shrewd when it came to money. There was Betty, his youngest daughter, who was only ten and still at school and very much his financial responsibility.

'Ciss' Johnson had been quick to realise how important a career in aviation had come to mean to Amy, and as such was willing to work on persuading her husband to set aside his genuine misgivings, and help their daughter to follow her chosen path. Then one sultry hot July weekend the Johnson family were struck by a personal tragedy that hit them all very hard. It was this event which perversely proved the catalyst, and forced Will Johnson to change his mind.

The last Saturday in July 1929 was the day that changed life for Amy and her family forever and the first tragedy to befall Will and 'Ciss' Johnson, compelling them to confront the reality that no parent should have to face: the loss of a child.

It was the Shell-Mex open day, and as a diligent and loyal employee, Teddy Pocock felt obliged to attend. He had tried in vain to cajole Irene into going with him. He shouted goodbye to her and set off alone. Mollie visited her sister later and stayed for a while, leaving with no indication of the tragic circumstances that were to follow.

It was Florence Carr, the Pococks' housemaid, who, letting herself into the house later that evening, discovered Irene. There was an eeriness about the place and a sense of something not being right which pervaded her mind, but the strong smell of gas which alerted her to Irene's fate. With the help of a neighbour, they discovered Irene slumped over the gas-fired oven, with her head to one side resting on a cushion. She was rushed to Hull Royal Infirmary, but it was too late. She was declared dead. Teddy Pocock, returning from the company Open Day, was given the shocking news at the Paragon railway station later that evening that *he* was a widower.

Will returned from his business trip to Scotland and, having collected 'Ciss' and Betty and determined to have all his family around him, sent a telegram to Amy in London. Amy returned home to Hull, deeply shocked and unable to fully comprehend the reality of what had happened. The younger sibling of whom she had once been so very protective was gone. Taken by her own hand, a loss so agonisingly real and such a tragic waste.

The first ordeal facing the Johnson family, was an inquest. For a suicide in the home it was routine, and the Coroner, J. Divine, was an old family friend who had known Teddy since he was a boy; he was as kind and as compassionate as his position allowed. He read the suicide note that Irene had left, and then handed it to Teddy, saying that its contents need never enter the public domain. Will mentioned Irene's earlier suicide attempt, but added that the last time he'd seen his daughter she had appeared quite happy.

Coroner Divine recorded a verdict of suicide while of unsound mind. The stigma of suicide was hard for Will and 'Ciss' to bear as they could not apportion blame on anyone but their beloved Irene. The family's status within the city, due to Will's business acumen, only served to exacerbate their misery, and sense of isolation as people began quite cruelly to speculate on the possible motive behind Irene's suicide. Some even suggested that she'd been having an affair. Will took the family, along with Teddy, who had moved into the Park Avenue house, to Ilfracombe in Devon to escape the prying eyes and the vicious wagging tongues. Amy had to return to her job in London. She had stayed in Hull until after the service of cremation, but she was still struggling to come to terms with her sister's death.

Her emotions fluctuated wildly, sometimes veering on hysteria, and in her letters home she urged everyone to stay as cheerful as possible. 'Ciss' looked to her Methodist faith for some kind of solace, but nothing that her strict religious beliefs had ever taught her could fill an aching void that stayed with her forever. Amy had however vowed to make a major change in her life. She was now determined to abandon the law and pursue a career in aviation. For if the tragic loss of Irene who had once coveted her own ambitions to go on the professional stage or run off to Paris and live the bohemian life, had demonstrated anything at all to Amy, it was the fragile nature of life itself.

6

AUSTRALIA! A GEM OF AN IDEA

When Will Johnson finally acceded to Amy's pleas for financial assistance – albeit with deep misgivings lingering in his mind – it allowed Amy to tender her resignation at Crockers and endure a difficult notice period in a tense and uneasy atmosphere in which Vernon Wood, although genuinely sorry to be losing his personal assistant, was often forced to chastise her for the tardiness in her punctuality.

At the end of August, Amy and Winifred Irving, whose friendship and support had been more invaluable than ever to Amy in the month after Irene's tragic death, took themselves off to Ventnor on the Isle of Wight. It was a much needed break for Amy, whose grief was still very raw. Teddy Pocock had added to the Johnson family woes by leaving the Park Avenue house without warning, and Will saw his behaviour as a personal slight against him.

Amy also felt hurt and disillusioned by Teddy's conduct and was angry at the blatant ingratitude of it, especially after all that her parents had done for him over the years. In a letter to her mother, Amy laid all the blame at Teddy Pocock's door, adding somewhat obliquely that perhaps Irene had felt her decision was the only way out of an unhappy marriage.

Will told Amy that leaving Crockers was a very big step for her to take. It was a leap into the unknown, where her obvious enthusiasm for flying – although very much to her credit – was not on its own enough. She would have to focus her mind on making a success of her venture.

She left Crockers in the first week of September 1929 and in a determined effort to maximise her income, prepared to make a raft of radical changes, in her personal life. She moved lodgings again, to be within walking distance of Stag Lane, so negating the need for her own car, which she was able to sell for £40.

Will and 'Ciss' were doing all that they could to help Amy fulfill her ambitions as a pilot, and in purely fiscal terms, this amounted to a monthly allowance of £50. Amy had a mentor, and a keen ally in Jack Humphreys, but she still had her detractors. those who couldn't wait to spoil her plans for a career in aviation. Conspicuous among them, was her old nemesis Herbert Travers who resented the attention paid to her. Amy admitted in a letter to her father, that she had never liked Travers very much when he'd been teaching her to fly, and now that dislike had developed into a much stronger sense of personal loathing.

Now she had the club secretary, Harold Perrin to contend with as well, and like Travers, he was eager to cut her down to size. Luckily for Amy, however, Jack was a very experienced mechanic and as such he was invaluable to the structure at Stag Lane, and *he* looked upon the hangars as his personal domain.

As the days and weeks of a gentle colourful autumn slid ominously towards the long dark nights of winter, Amy discovered that she had little time spare for leisure. Her days were spent in the hangars; in overalls straddling the fuselage of an aeroplane, with her head bent over the engine, while the evenings were spent in her rented room in Elmwood Crescent, with her head stuck in a textbook, revising for the Ground Engineer's examination.

On the day that became known in history as Black Tuesday, 29th October, 1929, a profound sense of shock set in as news of the Wall Street Crash broke and gradually seeped into the collective conscience of the nation. Historians have suggested however tentatively that this was the day on which the gaiety and frivolity of the Roaring Twenties came to an end and according to F. Scott Fitzgerald the Jazz Age leaped to a spectacular death. A great many myths have grown with time about that fateful October day and bankers and speculators held their breath as the market haemorrhage continued well into November. Others were left in a stunned silence, to look back on the decade of gaiety and extravagance that had once been the perfect antidote to the horrors suffered during the Great War; the Great Depression which followed was seen as the inevitable price to pay for having enjoyed too much, for too long.

In November the date of her examination was set and before Christmas. *Licence No 31* was issued to Amy on 10th December, 1929. She was now the only woman in the world to hold a valid Ground Engineer's licence.

What Herbert Travers, whose own ambitions to attain his 'tickets' had been somewhat thwarted by Jack Humphreys' determination to mentor Amy, must have thought at this outcome is easy to imagine. To have the 'Platinum Blonde', a patronising nickname that had been attributed to Amy, whose abilities as a pilot Travers had so callously dismissed months before, beat him in getting her 'C' licence first must have made the seething Travers, more determined than ever to curb her ambitions.

Alas for Travers, Amy was winning new friends and admirers all the time within Stag Lane and from wider aviation circles. One such man whose respect she would win early in 1930 was Air Vice-Marshal Sir Sefton Brancker, whose enthusiasm and belief in the importance of civil aviation was absolutely unshakeable.

Another ally whose friendship Amy came to value was James Martin, or 'Jimmy', who hailed from County Down in Northern Ireland. Aeroplanes were his thing – although it was design and construction which interested him. If time and better luck had been on his side, Amy could so easily have flown to Australia in a James Martin aircraft rather than a second-hand *Gypsy Moth* biplane. Jimmy had wanted Amy as the test pilot for the aircraft that was meant to make his fortune and to establish Martin's Aircraft Works as a household name. Amy would often borrow Jack's Austin to drive to Denham in Buckinghamshire, where Jimmy had his factory. She would make the journey many times in the early months of 1930 to check on its progress when the idea to fly to Australia that was first planted as a tiny seed by Jimmy Martin began to grow in her mind. Unfortunately for him, Amy was now a girl in a hurry. While Jimmy did his very best to have his prototype plane ready and tested in time, his grandiose plans were just too slow getting off the ground.

In early January 1930 a reporter from the *Evening News* went to Stag Lane, looking for a story about a young girl flyer who'd recently passed an examination, thus earning her 'tickets' as a qualified Ground Engineer. Whether he expected to hear all about Amy's idea to fly to Australia in a newly designed aeroplane which was still under wraps and about which she couldn't say very much at all is however somewhat doubtful. It proved however that Amy had caught the bug. Flying to Australia was all that she could think about now, as though the idea had been at the back of her mind from the very start. Jimmy Martin had had his motives, and for him the potential success, and the publicity value of having a girl flyer make it to Australia in an aircraft that *he* had designed, was incalculable.

Although the plan never came off as Jimmy and Amy had envisaged it, he was gracious enough to acknowledge that Amy's success was still important to him. He stayed close to the project and was one of the select few who stood in the early morning mist on the tarmac at Croydon Airport that fateful day in May to wave Amy off and wait until her plane was just a small speck in the sky.

When the article finally appeared in the *Evening News*, Amy felt justified in venting her anger at some of its blatant inaccuracies. They had got her age wrong and they had claimed that she was already earning a living from aviation, whereas in fact her job in the hangars at Stag Lane was unofficial and unpaid. It was a point that Jack Humphreys had been keen to stress when Travers and the club secretary, Perrin, had objected to her presence there.

In some respects it was fortuitous that the reporter from the *Evening News* ever got the opportunity to interview Amy at all. It was Harold Perrin who pointed the reporter in Amy's direction, no doubt keen to gain some positive publicity for the club. It *was* significant, however, given Perrin's original coolness towards her. Amy didn't really care if the London Flying Club were exploiting her as a 'trophy' for their own ends, especially if the publicity that she attracted caught the attention of the right people whose financial support would prove crucial in turning her idea of flying to Australia into a reality.

Once her idea was in the public domain, Amy set about gaining the support that she needed in a frenzy of letter-writing. An early target for her efforts was the *United Empire Party*. She exploited the connections of Empire by stating in colourful prose, the merits of taking a message of unity from the Mother Country to people of the dominions in far-flung corners of the globe. She even ventured to suggest that her plane could be named, the *Spirit of England*, a slight variation on Lindbergh's *Spirit of St Louis* in which he'd made the groundbreaking solo crossing of the Atlantic in 1927.

Australia was an enormous challenge, but how Amy explained why *she* should be the one to carry the message of the Mother Country is unclear. Her letter to Sir Thomas Polson made little impression on him or his Party no matter how much Amy gilded the lily in her prose.

She also wrote to the two great press barons of Fleet Street, Lord Rothermere at the *Daily Mail* and Lord Beaverbrook at the *Daily Express*. In the business of newspaper publishing these men were fierce rivals, but they had always shared a robust enthusiasm for developing the fledging industry of civil aviation. Besides using their papers as organs of opinion, they were also known to pay good money for any story of extraordinary feat in flying. For an unknown slip of a girl to attempt to attract the attention, let alone expect financial assistance, from these hard-headed entrepreneurs for such an audacious flight was brave, but also a little naïve. The timing was hardly ideal. The Depression was beginning to bite harder than ever: men were being thrown out of work on almost a daily basis; soup kitchens had become a harsh reality in towns and cities across the land and as hard-pressed charities gave what little help they could to the dispossessed and the needy, money was tighter than ever. While Amy gallantly argued that her flight could be a flag-waving morale boost for the country the £1,000 to £1,500, she had estimated she would need was in such austere times, a very big ask.

Amy also wrote to Sir Sefton Brancker in the flurry of letters she drafted during this period, as plans for her proposed flight gathered momentum. She had already marked 5th May as the ideal start date in her diary and as January slid towards February, time was of the essence. She had forgotten in her haste to sign the letter to Brancker. Luckily for her, the Air-Vice Marshal was sufficiently intrigued by its contents to do the necessary detective work and track her down, even though Amy had fortuitously mentioned her intentions in a recently published article in *Flight* magazine. Brancker managed to trace Amy to Stag Lane where he kept his own *Moth*. He took pains to point out her elementary error, in assuming that every titled person was inevitably endowed with surplus cash.

He was willing to make the necessary overture to Lord Wakefield of Castrol Oils to whom Amy had already written without success. Lord Charles Wakefield was as renowned for his great acts of philanthropy as he was for his strict adherence to Christian beliefs and while his support for the various feats of heroism that he espoused were crucial to the beneficiary – Castrol Oils had been integral to Alcock and Brown on their 1919 crossing of the Atlantic – to Lord Wakefield they often proved to be the act of a very shrewd business brain. So Wakefield considered very carefully whether to respond to Amy's letter asking for his support. Even the positive influence of Sir Sefton Brancker, or the benign opposition of her nemesis Herbert Travers alone were not enough to sway him either way.

In spite of this period of uncertainty, while she waited for Lord Wakefield to respond Amy continued to make what preparations she could for the epic journey that she was still determined to undertake. She was acutely aware of the necessity for setting off in early May to miss the worse of the monsoon season or put off her flight indefinitely and run the risk of another aspiring flyer stealing her thunder by reaching Australia before her. Bert Hinkler, whose own 1928 record still stood, had allowed rumours of his plans to seep into the public domain and there were reports that he had a new plane under wraps at his home near Southampton. While there was some news value of him making an attempt at bettering his record the potential merit of it being broken by a girl flyer was that much greater.

Lord Wakefield finally agreed to meet Amy at his offices on the Wednesday after Easter and barely three weeks before 5th May to assess the character of the girl whose abilities as a pilot he'd had already been appraised of by the latest champion of her cause, Sir Sefton Brancker. Wakefield wanted to see if she had the mettle, the sheer bloody-mindedness and guts to get to Australia. So a slightly nervous Amy sat in front of a large imposing mahogany desk, behind which sat an equally imposing man in whose philanthropic hands her fate now rested.

Lord Wakefield told Amy that he was broadly supportive of her plans and that he would demonstrate that support in purely fiscal terms by pledging half towards the cost of an aircraft and using his vast network of connections across the globe to provide Amy with the fuel she needed on the various stops along the route. There was however one strict proviso: that she agreed to do a course in navigation before setting off – no mean feat in the short time available.

Amy knew this was the benign influence of Major Travers at work, albeit from a distance. Although a little piqued that he could still exert his influence over her, she adhered to the condition as a small price for getting the vital support she needed. She warmly expressed her appreciation to Lord Wakefield; as did Will Johnson writing to thank the benevolent peer for the support and the faith that he had shown in Amy. Will himself pledged £550 to Amy for the trip, a substantial gesture, set against the dire economic conditions at the time.

Amy set quickly to the task of acquiring an aircraft to take her to Australia. She wanted to avoid relying on a *Moth*, as they were so well established now. There wasn't any novelty value to whet the appetite of the press, while De Havilland no longer needed to rely solely on the oxygen of publicity that was generated by a record breaking flight. With the Martin prototype now out of the running, Amy considered looking abroad although conscious of the fact that she would risk losing the kudos of her flight being a British attempt. It appeared when everything had been considered a second-hand *Moth* was her best option from the economic and the publicity perspective.

The *Gypsy Moth* that became *Jason* and forever identified as Amy Johnson's plane had belonged to Captain W. L. Hope.

It was a two-year-old *Moth* and after Amy had bought it for £600 she set about making it uniquely hers. Painted red and white with the registration letters GAAAH written large on the fuselage, Amy wanted it painted dark green; an inspired choice as emerald was the birthstone of her astrological sign – Cancer. It would also match the colour of her flying suit. The name *Jason*, painted with a flourish in silver, had been deliberately chosen. It was the telegraphic address of her father's business and was a fitting tribute from Amy to the father she owed so much.

Amy left *Jason* in the care of Jack Humphreys, who was to prepare the plane for the flight while Amy devoted her time to acquiring the provisions and equipment she needed, including a life jacket, spare propeller, which would be fastened to *Jason's* side and a revolver, which she had been advised was useful for keeping any unwanted visitors or curious plane gazers at a safe distance should she have to attempt a forced landing in some inhospitable terrain.

The short course in navigation had proved to be a valuable addition to her *curriculum vitae* of flying skills. Perversely, it helped to thaw the frosty relationship between herself and Travers. At the eleventh hour, he had decided that, in spite of all his initial misgivings about Amy's suitability, piloting skills, and her temperament for such a difficult flight – not to remain an obstacle to her ambitions. Amy realised that it was a big shift in position for him and demonstrated her appreciation accordingly. The weekend before departure came, perhaps a little too quickly for Amy, as she realised that she was now as ready as she could ever be for the great adventure of her life. If everyone is allotted one pivotal moment in the course of their life: then this was surely hers.

Will Johnson made the trip down to London on the Saturday, leaving the ever fragile 'Ciss' behind in Hull with their two younger daughters. Her nerves had barely recovered following the shock of Irene's tragic death. Amy and Will dined together at a London hotel and on Sunday she flew *Jason* from Stag Lane to Croydon Airport, leaving Will to make the journey by train. The bond between Will Johnson and his eldest daughter had always been very strong: he was determined to support his beloved Amy now. In the beckoning dusk of a spring afternoon, Amy brought *Jason* down on the rutted tarmac of Croydon Airport – the airport of Empire – as it was to become known would mark the starting point of so many of Amy's flying triumphs in the years to come. Yet for all the fanfare of the article, which had appeared in the *Evening News*, barely four months before, Amy's flight attracted little press attention. She had tried valiantly to sell her story.

The somewhat derisory sum of £25 had been bandied around which Amy would have accepted, but there were no takers. Amy would be grateful later that the newshounds of Fleet Street declined to take her story. For when the time came for the world to take notice of the Lone Girl Flyer her story would be worth considerably more than that.

Having put *Jason* in the hangar for the night, Amy retired to the Aerodrome Hotel. Will, whose room was along the corridor, woke her in the early hours before dawn on Monday, 5th May. She had endured a restless night from the constant noise of traffic. When her friend Winifred Irving broached the question of how she had slept, Amy snapped, 'Badly,' adding, 'The traffic kept me awake...'

A photographer stood on the tarmac later that morning and captured Amy standing in the cockpit of *Jason* in flying suit, helmet and goggles, looking tired and slightly pensive. He was the only one who had turned up to record the moment for posterity. It is ironic though that on the street from which the noise of traffic deprived Amy of sleep on the eve of the most important flight of her life, three months on, an expectant cheering crowd, of over a million stood for hours in the fading light of an August evening and hailed Amy Johnson as a national heroine.

7

CROYDON TO KARACHI

Amy reached Aspern aerodrome, fifteen miles from the centre of Vienna, the first stop on her route, touching down in the beckoning dusk of early evening, after nearly ten hours in the sky. Her delayed take-off from Croydon had been due to the nauseating smell of petrol from a leaking connection, and then a veil of fog hanging stubbornly over the Channel threw her plans into disarray.

The engineers would not let Amy do anything on landing, supervising the overhaul and care of *Jason* themselves. They were very friendly towards her, but also typically Teutonic in their efficiency, leaving Amy to dance hither and thither around them as she anxiously tried to ensure the checks were done to her specification and although she was fiercely protective of *Jason,* she finally surrendered to their will, leaving them to do their job.

The centre of Vienna was too far away for her to go in search of a plush hotel, so when she discovered there was no suitable accommodation close by Amy knew she was at the mercy and kindness of strangers in a foreign land. A friendly caretaker, took pity on her, lending her his bed for the night. Amy was up very early next morning eager to be on her way to the second stop on her flight. It was a journey that would take her to the outer frontier of Europe in Constantinople, the city on the Bosporus, once at the heart of the mighty Byzantine empire. The city that is known today as Istanbul was a teeming overcrowded metropolis of golden minarets, and aromatic bazaars where Europe and Asia converged at the gateway to the East. Amy had been made acutely aware of the problems that she was likely to experience in Constantinople. There was no sense of cordiale felt towards Britain; in many respects resentment simmered openly. The once great Ottoman Empire, the former rulers of Turkey, had aligned themselves to Germany during the Great War, and in defeat they had suffered accordingly. The fastidious and overzealous officials were notorious for delaying foreign pilots deliberately, going over their papers piecemeal, checking and rechecking that every minute detail was in order.

Amy hadn't had the luxury of time to waste in getting her papers authorised at the Turkish Embassy in London, and so she had set off with nothing more than letters of introduction from Sir Sefton Brancker to help oil the wheels of steadfast Turkish diplomacy.

Amy had covered the 800 miles according to her plan, by landing at San Stefano aerodrome before sunset; a feat which she hadn't quite managed on the first leg between Croydon and Vienna, when robust headwinds had reduced *Jason's* speed. After enduring an hour and half of inspection and a long, very frustrating wait at the hands of Turkish officials, darkness was rapidly encroaching, and she still had to change the oil and complete other routine checks, having promised Jack Humphreys that she would do a complete overhaul at each stop along the route.

Amy spent three hours with the aid of her torch and from the light of a car's headlamps overhauling *Jason* and putting him to bed for the night. So she was virtually dead on her feet with fatigue, when checking into her plain and spartan room at a nearby hotel. She found a telegram message which forged a poignant link with her past in the shape of best wishes from her former lover Hans Arregger.

Although Amy felt that the letters of introduction, which she carried from Brancker, had helped her in dealing with the Turkish officials, she was by no means confident that, given their reputation for awkwardness, they would allow her to fly out of the country without the necessary exit permit, merely on the strength of them. Her plan, once permission had been granted, was to make the long 'hop' towards Baghdad, the ancient city of the Caliphs. Steeped in history, the capital of modern Iraq had its fair share of souvenir hunters who would rip a plane to bits should a pilot be forced to land in its inhospitable terrain.

Amy had written to Bert Hinkler for his advice, on various places along the route – a somewhat audacious step, given that she had his 1928 record firmly in her sights.

Luckily a spirit of sporting goodwill existed between the pioneer flyers of the day and the diminutive Australian advised her as best he could. One of Amy's greatest fears was landing in the desert to be set upon by a band of natives; it was for this reason, that she had packed a revolver into her luggage – albeit without much thought as to whether she'd manage to use it.

As the morning wore on, with no sign of permission to take-off being granted Amy grew increasingly frustrated and angry. A leak in the fuel-tank, attended to by a friendly French mechanic, merely added to her weight of concerns, while inactivity forced her to abandon her plan to reach Baghdad in one hop, opting instead for a relatively shorter trip to Aleppo in French-mandated Syria.

Amy knew that to reach Mouslimie aerodrome at Aleppo, she would have to fly over the barren and hostile Taurus Mountains; a somewhat daunting task, even today. They proved a severe test of a pilot's nerve, and courage when flying a single-engine biplane with no radio, as Amy had in 1930. She had set her course to follow the railway track that wound its way through the deep ravines below, but dense cloud obscured her view, forcing her up to 10,000 ft to fly over the cloud mass. However the altitude upset *Jason*'s engine forcing her down again, and as the cloud gradually began to clear, she found herself heading straight for a sheer wall of rock.

Swift thinking on her part averted disaster, as she swerved *Jason* away just in time however the ordeal left her badly shaken, and although she now had brilliant sunshine to fly in, she was relieved to finally touch down at Mouslimie and put the worst leg of her journey thus far firmly behind her.

Amy landed weary, but reasonably satisfied with her day's flying at Mouslimie, in the searing heat of late afternoon. The French Air Force immediately went to work on *Jason* but demonstrated considerably more gallantry and respect towards Amy than the Austrian mechanics at Vienna, who had arrogantly dismissed her abilities as a ground engineer by virtue of the fact that she was female. She felt confident and happy to leave *Jason* in their capable hands, posing for photographs in her riding breeches, a man's shirt, and her Sidcot flying jacket which looked somewhat incongruous in the soaring heat. Amy felt that at Aleppo, the great adventure she had embarked upon was truly beginning to take shape; as she put miles between herself and the familiarity of home, she was more than ever on her own, forced to rely on her wits, courage, and sheer nerve. The next day she would head towards Baghdad, flying over endless miles of inhospitable desert in the searing heat.

Amy took off from Aleppo at dawn, with a distinct chill in the air, although she was aware that the temperature would rise considerably as the morning wore on. She pursued the winding Euphrates river, which reflected up at her like a silver pool in the sunshine, as she negotiated the 500 miles separating

Aleppo from Baghdad, with the ever-present menace of violent sandstorms nagging at the back of her mind.

At 7,000 ft, she escaped the worse of the heat and the turbulence, but was reminded of how the higher altitude had upset *Jason's* engine the previous day. The atmosphere was hardly conducive to efficient text-book flying, but Amy knew she had to press on; the Hinkler record was beatable, but she couldn't afford the luxury of delays. She knew the extremes of weather were always going to be a factor that she had to work through, and in the desert sand storms were a common danger. *Jason'* suddenly lurched forward with a shuddering jolt, his nose dipping slightly. Amy held on tightly as she fell several thousand feet, trying hard to keep a cool head. Then for one awful moment, but what must have seemed like an eternity to Amy, the propeller stopped and she lost altitude again. The thermal eddies were jostling and tossing *Jason* like a small boat on a stormy sea.

Amy realised that flying against the storm was futile, and could be potentially fatal. If *Jason's* propeller had stopped *again* she couldn't be sure it would ever restart. She began to descend slowly, fighting to keep *Jason* steady as the sandstorm increased. She finally touched down in a blizzard of flying sand, tiny particles smearing her goggles as she struggled out of the cockpit, and began to cover *Jason* with a tarpaulin. As the wind gained momentum, she struggled to keep the cover from coming loose on one side before she had fastened it on the other. After almost half an hour of struggle, she had the plane covered, and the engine protected from flying sand. She took her luggage from the hold to use as improvised chocks under *Jason's* wheels, and feeling exhausted, she straddled the engine, with her revolver in hand to see off any unwelcome visitors waiting patiently for the ire of the storm to abate.

After a delay of almost three hours, alone in the desert, Amy was airborne again. In spite of her fears that the flying particles of sand would do severe, even irreparable damage to

Jason's engine, it started first time. She pointed her aircraft in a easterly direction, collected her belongings, and after one swing of the propeller, she was up and gaining altitude. Her plan was to follow the other great river which flowed into the ancient city, the Tigris, and soon the endless vista of reddish brown desert would be behind her.

When she finally came to land at Baghdad Amy suffered the first of many cruel setbacks which plagued her flight. There was damage, caused by her forced landing in the sandstorm, to *Jason*'s undercarriage, where one of the struts had sheared off on impact, so she landed at Baghdad with one wing down. This was the most serious setback thus far, and Amy was close to tears from fatigue and sheer frustration. Imperial Airways came to her rescue at Baghdad, as she was initially told that the spare strut would have to come from the De Havilland depot at Karachi some 1,500 miles away but from somewhere a new strut was found at the RAF aerodrome at nearby Hinaidi. Amy left the Imperial Airways mechanics to fix *Jason* while she took some much needed rest and refreshment. She then agreed to a tour of the fascinating city and its many ancient treasures, rather than stand around on the tarmac, constantly checking the time and allowing her anxiety level to rise.

An overnight stay in Baghdad had become inevitable, and Amy woke next morning anxious to see her beloved *Jason* and found him ready for the next leg of their epic flight. The new strut had been fixed, and the engine was completely overhauled. Amy was relieved and very grateful. In borrowed shorts, for which she had discarded her Sidcot suit, she climbed into the cockpit, and took off, heading towards Bandar Abbas, a tiny and remote fishing port at the top of the Persian Gulf in what is modern Iran. On the world atlases of 1930 it was still known as the kingdom of Persia, and its ancient rulers, the Shahs, had for centuries reigned supreme on the Peacock Throne.

Amy had chosen Bandar Abbas as a halfway point between

Baghdad and Karachi. It was a remote and uninspiring place about which little was known, then or since, but when Amy landed there on the fifth day out of Croydon – Friday, 9th May – and discovered to her heartbreaking horror that the bolt to secure the new strut had broken off, she could have been forgiven for thinking that the tiny fishing port was about to earn itself enduring fame as the place where her gallant stab at Hinkler's record finally came to an end. She had absolutely no faith that the good fortune she had enjoyed at Baghdad could be repeated in a place as remote and small as this. Her sense of despair was almost tangible. Yet she knew girlish tears would get her nowhere. 'A Lone Girl Fyer' was enough of a novelty to begin with and her pluck had done her considerable credit so far. Nonetheless she had a splitting headache, and was in desperate need of a hot bath while weariness and total fatigue added to her woes. She approached the house she had seen from *Jason's* cockpit, and discovered it was the home of the British Consulate, a chap named Richardson, who explained to Amy that the airfield was no longer in regular use hence its poor state of repair.

A 'white knight' in the shape of a young mechanic named David, would, she was informed, take a look at *Jason* and see what he could do to fix him. Meanwhile Amy was taken into the Consulate's house, introduced to his wife and daughter, and despite her tiredness was encouraged to enjoy their considerable hospitality. Amy took full advantage of the rest and sustenance offered her by the Consulate and his wife. She was grateful for the headache powder, but her pressing concern for *Jason* preoccupied her. The obvious affection she had for her plane was palpable. She went out to the desolate airfield on which she had been forced to abandon him and to her delight she found him back on both legs.

The young mechanic David explained he'd found the new bolt among the spares he had left over from the RAF who'd once used the now derelict airfield. David and an inquisitive group of locals, gathered round to satisfy their curiosity about the girl pilot who'd just flown into their lives.

Amy now had the task of draining off the old oil and replacing it with fresh Castrol XXL but, inevitably, tracking down the supply was yet another time-consuming ordeal. Thus it was nearly 2.30 am before she *finally* got to bed. The humidity level had dropped considerably and there was a cool breeze in the air, and although good fortune had smiled on her yet again – this time in the form of David thus averting another potential disaster – the precious time she was losing at each stop, gnawed at her. She simply could not afford these delays and was acutely aware that she had to make real progress to avoid the worse of the monsoon season. Amy left Bandar Abbas on 10th May en route for Karachi – then the northwest frontier into India, which in 1930 was still part of the Empire, albeit with a thirst for independence, and civil unrest ripe in the air. At this point people back home began taking notice; Amy was longer to be casually dismissed as some silly girl who had set off on a great adventure with just a thermos flask of coffee and some boiled sweets.

She landed at Karachi in temperatures exceeding a hundred degrees, and was bombarded with bouquets of flowers as cheering crowds rushed forward just to touch her. Garlands were hung around her neck. *The Daughter of the Empire* had arrived. Bert Hinkler had only managed to reach Basra on the Persian Gulf on the sixth day of his record making trip, so Amy, the former typist from Hull, had earned an accolade that was hers alone. Whatever else she may achieve, she had set a new England-to-India record!

A fierce bidding war quickly ensued between the press barons of Fleet Street each of them now willing to pay handsomely to secure Amy's story for *their* newspaper. The editors, who hadn't wanted the story, even for a derisory

£25 a few weeks before, were now printing banner headlines proclaiming her: *The British Girl Lindbergh!*

A newly acquired celebrity status allowed Amy to stay at Government House in the Karachi Municipality on her own terms. Pleading, overwhelming fatigue, and a lack of formal attire mercifully allowed her to escape the official banquet, they had planned to hold in her honour. Instead she tried to catch up on some much needed sleep. Back home in England, a media storm was gaining momentum around her.

8

THE LONE GIRL FLYER

Amy left Karachi on 11th May. While she slept beneath cool sheets in her room at Government House – the most restful night's sleep that she had enjoyed since leaving Croydon – the mechanics set to work on *Jason*. The engine was completely overhauled, and the tools that she had lost in the Iraqi desert were replaced. They also discovered that the enthusiastic, but largely inexperienced, mechanic David had made an elementary mistake in fixing an extra washer onto *Jason*'s plugs, which if it had been on the carburretor side could easily have spelt disaster.

Out of Karachi, en route for Allahabad, Amy was in a euphoric mood. She was now two days ahead of Hinkler's record, but as yet blissfully unaware of the media frenzy that her achievement had provoked at home. Her next stopover stood at the confluence of two great rivers: the mystical Ganges and the Yumana. These vast bodies of water were like a scenic oasis after the black featureless landscape of the Sind desert, which had stretched endlessly beneath her out of Karachi.

She landed in an airfield, that she believed to be Allahabad thinking she had made exceptionally good time, but quickly discovered that she had landed a few hundred miles short of her destination at Jhansi, a small sleepy town, which basked in suffocating heat. Amy quickly realised that she hadn't enough fuel to reach Allahabad.

Having taken off from Jhansi, following directions given to her by a man on the ground she had no choice but to turn back towards Jhansi and suffer yet another infuriating unplanned delay.

After the euphoria of Karachi, she could have been forgiven for thinking that the good fortune, she had enjoyed so far was gradually slipping away. She struggled to land *Jason* with an unnerving sense that her plane was out of her control. As she touched down on the parade ground of the third and eighth Punjabi Regiment, the aircraft crashed into their barracks, and damaged one wing. Having scattered left and right, out of the path of the careering biplane the soldiers ran to offer whatever help they could to the pilot and were astonished to find a young woman dressed in men's attire emerge visibly shaken, but unharmed, from the cockpit.

A bearded Indian mystic with the skills of a craftsman repaired *Jason* to the extent that he was almost as good as new, and in better shape than when they had left Croydon. Petrol was found at an aerodrome ten miles away from where Amy had force-landed and was brought by car. Amy then set about the meticulous task of overhauling the engine while onlookers watched in awe as the young girl pilot performed the task of ground engineer with skill and minute attention to detail.

They offered her their help, which she politely declined, until they could see that she was virtually dead on her feet, at which point they insisted that they finish the work while she lay down and supervised them from a distance. Amy was offered a hot bath in the colonel's house followed by dinner, after which she regaled them with snippets of her journey to date.

Amy found that sleep eluded her at Jhansi due to the suffocating heat and even though the temperature had fallen slightly by nightfall it was still unbearably hot. Just as the noise of traffic outside the Aerodrome Hotel had deprived her of sleep, on the eve of her departure from Croydon, Amy later remarked that, *Jhansi was hotter than I ever imagined a place on earth could be*.

She left the pleasant little town at dawn the next morning. Her plan was to reach Allahabad for refuelling, and then onto Calcutta. This would be her last stop in India before she left the 'Jewel of the Empire' behind her to head off across the Bay of Bengal, towards Rangoon, and into the heart of the jungles of the tropical Far East, where the monsoon season, which she had always known would be the severest test of her piloting skills, would soon become reality.

Amy landed at Calcutta, and found a telegram from her parents awaiting her. Will Johnson had become something of a 'media star' thanks to Amy's achievement of setting a new record to India.

Trying valiantly to protect the ever fragile 'Ciss' he took all the media attention upon himself, and although he was not then adept at playing the media game he quickly adapted to their constantly demanding presence in his life, he was able to deflect their intrusive questions, relating to private family matters whenever it suited him. He was immensely proud of what his beloved Amy had achieved thus far, and was delighted albeit somewhat surprised to receive a standing ovation from his fellow Rotarians at a convention in Edinburgh. Meanwhile the clamour to secure the exclusive rights to Amy's story was spiralling out of control the figures being quoted staggered belief. Each newspaper was eager to score a coup by securing Amy's story before she touched down on Australian soil. When the figure of £2,000 was quoted Will Johnson could barely believe what he was hearing; this was beyond the comprehension of the average man in the street. Amy eventually received five times that amount for the exclusive rights to her story and a nationwide lecture tour. Lord Rothermere had held his nerve, against his competitors, and won the prize: triumphantly securing Amy for the *Daily Mail*.

Although the money would attract some adverse publicity, inciting one Australian newspaper to accuse her of greed – albeit in subtly veiled terms – the public were generally supportive. Will and 'Ciss' Johnson had come to London to negotiate the finer details with representatives of Associated Newspapers. Later, together with Jimmy Martin, they went to a plush West End hotel, and celebrated the fantastic deal that they had struck on Amy's behalf. Amy was advised on taking off from Dum Dum aerodrome at Calcutta that she was likely to encounter tricky weather conditions virtually all the way to Rangoon. This was the leg of her journey, which would prove the most challenging, requiring her utmost concentration, reserves of energy, and sheer mental stamina.

She had set her course to head south, hugging the marshy area of the Bay of Bengal, to which the mystical Ganges stretched its fingers furthest, and thence to the Irrawaddy Valley. The tricky weather proved with an unnerving accuracy to be every bit as bad as those on the ground at Calcutta had warned her it could be. The rain became so intense that it obscured her view, from, *Jason*'s cockpit and in these extreme circumstances, Amy allowed an atmosphere of panic, to set in, and began to doubt her navigational abilities.

She flew over Akyab close to the northern coast of Burma, but she soon discovered that the aerodrome there the only serviceable one lying between Calcutta and Rangoon was completely flooded, and so the chance of a short respite from the torrential rain although it would mean forcing upon herself yet another unscheduled delay that she could ill-afford, was now lost to her.

She had no choice but to press onto Rangoon and hope that the rain would ease off. Amy flew so low over the railway station, that *Jason* almost touched the roofs below. It was a somewhat daunting experience, for those on the ground sheltering under their umbrellas. The rain stung Amy's eyes as she tried desperately to focus on the various landmarks dotted around below her.

Her intention was to land on the racecourse, but as she headed for the wide open ground that she believed was her chosen landing ground, she saw people waving at her frantically, and not as Amy thought in greeting but as a warning that the raceourse was further on. Amy discovered her error with a thunderous thump as she brought *Jason* down on the football pitch, too late to have aborted her landing, due to a low hanging clump of trees. She careered down into a trench, the post puncturing *Jason's* left wing and ripping a tyre wide open.

Amy jumped from the cockpit, and seeing the extent of the damage her error had inflicted on poor *Jason* her emotions gave way under the stress and strain of the day. Through a torrent of tears she sensed this setback could well spell the end of her record-breaking flight. While nobody would have blamed Amy for abandoning her dream of reaching Australia, at that moment, her hopes for beating the Hinkler record had now diminished considerably. Amy discovered after the initial shock of her crash landing and once the tears of despair, had subsided, that she had come down in a sports field five miles short of Rangoon, at a place called Insein which somewhat fortuitously was the home of the Government Technical Institute.

Amy wasn't in the right frame of mind to appreciate any sense of good fortune; for the sight of the mangled *Jason* with its propeller broken was more than she could bear. Closer inspection revealed that a strut on the undercarriage had snapped off together with the damaged wing and the ripped tyre that she had witnessed on landing. The damage was as bad as Amy had feared, and yet another long delay while she waited for the spares to arrive, seemed inevitable.

While the staff and students at the Technical Institute were quietly confident that they could repair *Jason*, Amy didn't share their optimism, and was acutely aware that, had such an accident occurred in England, it would have kept her grounded for at least a week. This was an agonising time for Amy, as she had little choice but to place herself entirely in their hands and hope they could match their obvious enthusiasm and willingness to help with the necessary skills to fix *Jason* and get her airborne again.

The damaged wing was carefully removed, and carried away for repair; meanwhile Amy unstrapped the spare propeller, which she had flown with all the way from Croydon and fastened it in place. She then set about cleaning the engine, supervising the students, in stripping the mangled metal section of *Jason* which was to be straightened wherever possible and welded together. An English forestry inspector who rode like a gallant knight to Amy's rescue instructed the students in their allotted tasks while he volunteered his help in getting the damaged wing repaired. When Will Johnson learnt of Amy's crash-landing in Burma, he remained stoically confident, telling the ever inquisitive reporters, that he believed Amy could endure this setback, and much more in her quest to reach Australia.

The forestry inspector managed to make new wooden ribs for *Jason*'s wing but the question of what could be used as substitute material remained as most of Amy's linen supply had been used for running repairs at Jhansi. An inspired moment of genius provided them with a solution by way of the hard-wearing shirts worn by the local men. They had been made from surplus aeroplane fabric sold off at discount after the Great War. Now the fabric would be used for its original purpose as it was torn into strips, and became part of *Jason*'s new wing.

Amy's achievement in reaching Australia, due largely to her extraordinary determination to realise her dream and sheer *pluck* in the face of so many adversities, was realised also because of the generosity of ordinary people that she met along the route; whose curiosity and fascination in the Lone Girl Flyer compelled them to offer their assistance wherever it was needed, and never more graphically so than at Insein. Even the local fire engine was commandeered to carry *Jason* to the racecourse that Amy had been searching for in the driving rain when she had crashed, and would now mark her take-off point out of Burma.

It was her eleventh day out of Croydon, and in 1928 Bert Hinkler had reached the same point on the eleventh day of his epic flight. Amy had lost the two-day lead she earned amid the hurrah of her recording-setting arrival at Karachi, and her bid to beat the record of the diminutive Australian was finely balanced. Amy was understandably nervous about taking off from the racecourse at dawn on Friday, 16th May. She allowed herself some peace of mind by taking *Jason* up for a short test flight after which she set off for Bangkok, capital of the Far East tourist haven that is modern day Thailand, which in 1930 was marked on world atlases under its old name: the kingdom of Siam. A large crowd of cheering Burmese saw Amy off as *Jason* with took to the skies once more, and in a vigorous wind, Amy took him to 9,000 ft to fly above the cloud mass, crossing the Gulf of Martaban, in conditions as tricky as those that she had endured on her approach to Rangoon two days before. Tiredness and an irritating headache gnawed at Amy's concentration, and she landed at the wrong end of the Don Muang aerodrome, some twenty miles from the centre of Bangkok, amid a crowd of cheering Siamese, who as eager to satisfy their curiosity about the Lone Girl Flyer as the Burmese had been. Amy was as ever grateful for the welcome that she received on the ground which helped to dissipate the long, lonely hours that she spent in the cockpit, reliant only on her wits.

Time was of the *absolute* essence: a priceless commodity she couldn't afford to waste. Sightseers, were as eager as ever to talk and touch her however well intentioned, they were an inevitable strain on that commodity. Transport to the centre of Bangkok was generously offered, to which Amy politely and graciously declined.

She needed to overhaul the engine. As night fell and the temperature dropped the inevitable plague of mosquitoes became a nuisance to Amy, and her small team of eager Siamese mechanics. Later Amy was forced to undress in the dark, and then beat the pests from her bed, before settling down for some much needed sleep. Amy woke early on 17th May to leave Bangkok at dawn, but once again fate conspired against her. Her plan to cover the 900 miles from Bangkok to Singapore in a day's flying was entirely feasible yet the elements, and a problem with the engine cowling, which flew open shortly after take-off, forced her back to Don Muang to get the cowling fixed and to watch in sheer frustration as precious minutes ticked by.

Heavy rain was another hazard impeding her progress forcing her to fly as low as she dared. But after more than eight hours in the air, and less than half her planned distance covered, she knew that reaching Singapore in a day was no longer possible. Instead she opted to land at Singora, some 400 miles short of her destination.

Back home a news hungry public, fascinated by the heroic exploits of the Lone Girl Flyer, waited for the latest snippet of news on Amy's progress to Australia. The *Daily Mail* boldly informed its readership that it had secured Amy's story for their paper. There was however a flip side to the burgeoning sense of optimism which wrought a collective groan of disappointment from those urging Amy on to set a new record when the realisation hit that on his thirteenth day in 1928 Bert Hinkler had made it to Singapore.

This fact didn't seem to matter, as such was the sense of national pride in what Amy had achieved thus far that the full weight of public goodwill remained with her. Amy took off from Singora using the road as a makeshift runway rather than risk taking off on sand, so soft that it could barely take the weight of *Jason*'s wheels. The excited crowd of smiling Siamese stood on either side of the road several lines deep; such was the novelty value to them of a Lone Girl Flyer.

It made the manouevre of her *Moth* that more difficult for fear of hitting her avid spectators as she gained the necessary speed for take-off with a full tank. Amy urged and pleaded with them for their own safety to allow her as much space as possible. Although landing had always caused her more concern, this was her trickiest take-off thus far, surrounded by onlookers with no appreciation of the potential danger facing them, as Amy forced *Jason* into the air, en route to Singapore, then turning to wave down at a crowd of awe-struck people gazing upwards at an ever diminishing speck in the sky.

Amy's arrival at Seletar aerodrome in Singapore, another outpost of the Empire, and the rigid formality of her welcome, as a daughter of the Mother Country was in stark contrast to the overenthusiastic and warmly spontaneous greeting that she had received at Bangkok and Singora. Men stood on the tarmac in their starched white uniforms, together with their women in formal dresses, holding floral-patterned parasols, that would not have looked out of place at the summer garden party of an English country house.

With pomp and circumstance, however muted, they awaited the arrival of the aviatrix for which they had been forewarned, seeing her as a welcome distraction from their strict observance of the Sabbath which barely differed from one week to the next.

Amy was pleased to have a cable from her father waiting for her at Singapore. Any link with home for the weary pilot was welcome and reassuring at that stage of a long distance flight. Its contents however were well less encouraging, coming at a time when Amy, *so* close to exhaustion, could have done with any kind of morale boost. He informed her that to beat Hinkler's record she would have to reach landfall at Darwin, on Tuesday, 20th May. Amy realised that with the reduced speeds and the distances covered at the end of each day's flying this feat was now beyond her. Will still urged caution. Although her spirit remained strong, her endurance was being severely tested and the extent of fatigue showed behind her eyes.

Even the prospect of her beating Hinkler's record for overall hours in the air was quickly diminishing. However bleak it may have seemed, giving up now was unthinkable. Reaching Australia *and* being the first woman to achieve it solo had become her dream. So she simply *had* to go on.

Amy set off for Surabaya in Indonesia: a journey of 1,000 miles, which she wanted to cover in a single day, although mindful of the fact that she had set herself such a target many times since leaving Croydon, only for the elements to thwart her ambitions.

She arrived at Surabaya in the Dutch East Indies mentally and physically shattered. The monsoon rains and dense low cloud had come once more to hinder her progress and Amy suddenly felt compelled to pray. She had taken everything the weather had found to throw at her with fortitude and with as much good humour she could manage, in her sleep-deprived state. Now she was hoping for help from the Almighty. Despite her strict Methodist upbringing Amy didn't consider herself to be particularly religious, although she'd known that ever since Croydon her mother had been praying for her guidance and safe return.

Another delay at Java meant that the Hinkler record had finally gone. In February 1928 he had touched down on Australian soil on his sixteenth day. This was her seventeenth. Amy had known time was ticking against her, ever since the crash-landing at Insein. Now she had it *confirmed.* She would still become the first woman to fly solo to Australia, but as far as record-breaking was concerned, she would have to settle for England to India.

Betty Johnson, now aged eleven, and Mollie sent cables of congratulation to Amy stating how proud they were of their 'big' sister, as were the women of England. It was this perception of Amy's very ordinariness that so endeared her to her gender. From factory girls to waitresses, and salesgirls in the retail emporiums of the West End, they all felt that the former typist from Hull, was one of *them.*

On 22nd May, day eighteen, Amy set off for Atamboea, having benefited from a good night's sleep due to a delay forced upon her by a problem with the magneto on *Jason* keeping her at Surabaya longer than she had intended. Ahead lay another 1,000 miles to Atamboea. It was here that England took a collective deep breath and a dark pall of anxiety loomed menacingly, when, having been seen shortly after noon, Amy was feared missing. Anxiety levels rose the longer time went by without firm news of her, and a sense of numbness fell on the Johnson family home, as they waited patiently, hoping *and* praying that their very worst fears would be assuaged.

The nation was suddenly paralysed with concern for Amy's safety. Her exploits had captured the imagination and the *respect* of a nation ever since her record-setting arrival in India. With no firm communication links with Atamboea, newspaper editors were caught in a dilemma of having to force their readers to confront their worst fears in print: on 23rd May newspaper billboards carried the alarming headline "FLYING GIRL MISSING!" The nation awoke to see these headlines, and gasp; then to wait and *hope* some more. At Surabaya the Dutch had two seaplanes on standby to attempt a search. This was the last denouement, that an adoring public could ever have contemplated for Amy, and the sense of disbelief was almost tangible. 'Ciss' Johnson as ever acutely aware of the tragic events less than a year before, was sick with worry. When Amy was finally sighted and confirmed as safe Will told a reporter: *'he wouldn't want to go through another night like that for a fortune'.*

Amy set off from Timor on 24th May, a Saturday, unaware of the stir that her supposed disappearance had caused back home. Ahead of her lay the shimmering blue stretch of the Timor Sea:

shark-infested waters that were particularly dangerous as the long-distance pilot confronted them at their most exhausted and therefore liable to make mistakes. For Amy, after all the ill-luck and tricky weather that had *so* bedevilled her record breaking attempt since Karachi this last leg of her epic journey proved to be one of the most mundane, but for all that it required her utmost concentration.

Shell had arranged for the oil tanker *Phorus* to stand guard for her midway across the Timor Sea. In the mid-afternoon, when the cloud broke and the sun shone through, Amy caught her first glimpse of land. Australia! At this stage, so close now to what had seemed so far away for so long Amy was overwrought with emotion.

At Fanny Bay aerodrome a large crowd had began gathering to welcome the Lone Girl Flyer who had come from the heart of the Empire to see them, and on Empire Day itself. It was an irony that could not have been lost on Amy, given her overtures to Sir Thomas Polson of the United Empire Party to garner support for her flight with just that angle in mind.

As Amy brought *Jason* down on the pot-holed surface of the aerodrome, the crowd broke out in cheers and spontaneous applause. A lone police officer was charged with keeping order but his efforts were hopelessly in vain. As people rushed towards the *Moth* Amy was in tears. Throwing off her goggles, she rose tentatively from the cockpit to greet her public. Back home as news of Amy's safe arrival fizzled down the news wires her success was hailed with relief and joy bordering on the euphoric.

Newspaper editors who twenty-four hours earlier had been forced to report the grim news of her disappearance now hailed her achievement with a bold two word headline: 'SHE'S THERE!' Such was the extent of Amy's fame now, this would suffice, while on street corners across the land news vendors could shout her name with pride.

AMY IN COCKPIT OF "JASON"

9

AUSTRALIA: A WHIRLWIND TOUR

"Amy! Wonderful Amy!" These words which formed the lyrics to the Jos Gilbert song was a sentiment shared by a nation in mid-May 1930. But for Will and 'Ciss' there was quite simply relief in the knowledge, that their beloved Amy was finally there and that she was safe. King George V and Queen Mary sent telegrams of congratulation and already there was clamour for Amy to be formally honoured in some way. The title Dame of the British Empire was muted and yet this, the highest honour that could be bestowed upon a woman not yet twenty-seven, was considered too old and stuffy for a vibrant courageous girl-flyer like Amy.

At Karachi she had declined an invitation to a formal dinner at Government House on the grounds that she had no formal attire. Such a lame excuse would not be allowed to spoil her first night on Australian soil. So Amy in the presence of the government representative and other dignitaries was wined, dined, toasted and lauded, until finally she was allowed to retire for the night where she slept for eleven hours.

The first reality Amy had to come to terms with was her new found celebrity status, thrust upon her, during the course of nineteen gruelling days she had spent in the air and on the ground, insulated from the world around her and focussed only on realising her objective. Now she was seen as public property. Suddenly her life didn't feel like it belonged to her anymore and a tangible resentment began to fester inside her.

A horticulturist named the Amy Johnson Rose after her, and formal luncheon invitations for Amy to be the guest of honour at one association after another abounded. Five days after her arrival, Will, 'Ciss' and the girls, finally got to speak to Amy, but what should have been a private family reunion was deemed a photo opportunity.

The *Daily Mail* were determined to protect the investment they had made in Amy, and Will urged her not to say anything to any rival publication that might jeopardise the lucrative deal that had been struck on her behalf. Castrol Oils whose titular head Lord Wakefield had pledged 50% of the cost of buying *Jason* now used his considerable business acumen by exploiting the faith he claimed that he always had in Amy for the benefit of his company.

Their man in Australia shadowed Amy wherever she went, and showed no hesitation and sometimes spared no mercy in reminding her of her obligations to Castrol Oils, while their great rivals Shell also vied for their share of the spoils.

Each day over the next three weeks dawned with some civic reception, luncheon or dinner at which she sat at the top table and was expected to make a speech. It soon became apparent to Amy, and caused her considerable frustration, that the very purpose for which she had flown to Australia to attain the respect of the aviation industry for her abilities as a pilot, on merit regardless of gender, was getting lost amid an endless cavalcade of publicity which had her at its centre but over which she had little or no control. Amy quickly realised that she had paid a very heavy price for being the first woman to fly solo to Australia. She was just too famous now to secure the serious job in aviation that she had once so coveted.

The endless round of receptions and dinners began barely forty-eight hours after she had touched down on Australian soil. She was flying *Jason* from Darwin to the town of Daly Waters, and Charles Scott, a former RAF and Qantas pilot, was determined to set the pace flying a cabin biplane, which left Amy and the Shell plane struggling to keep up with him. Scott had aspirations as a record-breaking stunt flyer, and he possessed a somewhat chauvinistic attitude to female pilots in general; for Amy, he bore a particular resentment which he had no intention of concealing. It was a frustrating time for Amy as Scott knew the country well and by instinct the particular landmarks, having flown the route many times, but to Amy the scene below her was as unfamiliar as the desert near Baghdad. The outback consisted of sporadic small towns and endless dirt tracks. If Scott's sole intention was to highlight and exploit some perceived weakness in her abilities as a pilot, she was damned if she was going to give him the satisfaction.

The Shell plane could lag behind Scott in third place if they chose to, but Amy, with her teeth gritted, was determined to demonstrate some of the English 'pluck' which had brought her from the other side of the world. Charles Scott's attitude was *so* juvenile and typically misogynistic, but Amy knew that she could prove a point as readily as him.

By Tuesday 27th May the convoy of three aircraft had reached the Queensland town of Longreach which lay in the heart of sheep-shearing country. Amy was feted wherever she went in Australia; a fact which must have galled Scott considerably, as he was still setting a mean pace in the lead aircraft, determined to expose Amy as just a mediocre pilot who had been elevated to her celebrity status, largely due to a well organised but overblown media circus.

Whatever petty jealousy was motivating Scott, his efforts amounted to nothing. Amy was the star of the show, and the swelling number of people who came out to see her arrive in their small town, proved that point quite graphically. At Longreach four police officers led her to safety from a large overwhelming crowd. While Amy was more than able to cope with the petulant behaviour of Charles Scott, it wasn't the first time that she had encountered male chauvinism in its crudest form. Nor did she expect that it would be the last. A friendly face was always a welcome sight in a town of strangers, most of whom were avid autograph hunters.

Her latest knight was the De Havilland representative in Australia, Ian Grabowsky, who tentatively approached Amy to introduce himself. His friendly demeanour made an immediate impact on her, as she felt able to confide in him that she was suffering excruciating period pains. Despite the obvious delicacy of the situation, Grabowsky felt compelled to report this to Scott and Captain Stanley Bird, insisting that Amy should be allowed some time to rest. Captain Bird was the Castrol Oils representative in Australia and was flying along with an accompanying journalist as part of the entourage around Amy in Charles Scott's plane. Bird was a fully committed company man, somewhat unbending in his demeanour, and determined they adhere rigidly to the schedule. Scott clearly had his motive for being unsympathetic towards Amy; nonetheless Ian Grabowsky walked away, genuinely dismayed and disgusted by their ungallant response to Amy's obvious suffering. Amy was not surprised by their insensitivity but was grateful nonetheless to Grabowsky for his efforts on her behalf. Nor would this be the last time that he gladly assumed the role of her gallant knight before she left Australia.

When the convoy of planes flew out of Longreach heading towards Charleville Amy was still in considerable pain. As Scott and the Shell plane came into land at Charleville, the town's mayor and the welcoming party, fully expected the girl-flyer to emerge from the cockpit of *Jason* to a rapturous round of applause. However, when Amy failed to appear the looks on their faces was a mixture of shock and genuine disappointment. Bird was visibly seething at what he saw as a deliberate rebuff to the people of Charleville and as a blatant challenge to his authority. Scott however was typically unconcerned.

Amy had decided to take a detour towards Quilpie, another sheep-shearing town some 120 miles west where nobody was expecting her, and so there was no great fanfare heralding her arrival. She could simply enjoy the anonymity and have some much needed rest. Amy finally arrived at Charleville in the dwindling light of early evening to explain to the waiting pack of journalists, that her late arrival was due to her getting lost en route. Charles Scott was deeply sceptical of Amy's claim, while Bird, evidently embarrassed by her 'no show', was too angry to make any comment about what he perceived to be the increasingly petulant behaviour of his charge.

When Amy came into land at Eagle Farm aerodrome, Brisbane, a crowd of 20,000 people had gathered to see her arrive in the Queensland state capital, and thus be a part of history, and as such there was a tangible sense of anticipation and excitement, but none of them could possibly have imagined the unfolding drama that they were about to witness. Approaching Eagle Farm Amy misjudged her descent, overshooting the runway, and crashing into the boundary fence, turning her *Gypsy Moth* over on its head in a neighbouring cornfield.

Amy was suspended upside down, still strapped into her harness, but miraculously she escaped serious injury. Shocked spectators rushed towards the wreckage to offer any help they could, and saw Amy emerging tentatively from the mangled wreckage of *Jason*, visibly shaken by her ordeal. Whether this accident was due solely to fatigue and her excruciating period pains, or simply a random moment of poor piloting judgment, is unclear. She had developed a fear of large, uncontrolled crowds, of people massing around her and treating her as if she were public property. It was bad enough that Castrol Oils regarded her as though she were simply their asset. A trophy for them to do with as they wished, showing no regard whatsoever for her personal well-being and this combined with the thought of thousands of exuberant fans rushing towards her filled Amy with a sense of sheer panic.

Ever since she'd earned her 'wings' Amy had struggled to master the art of a 'textbook' landing but her experience at Brisbane, was by far the worst moment of her Australian tour. To see the 'plucky' girl from Hull, who had endured so much just making it to Australia emerge almost unscathed from such potential disaster, only served to endear her even more in the hearts of the Australian public.

Although deeply upset at seeing *Jason* carried off for yet more repairs from the cornfield adjoining Eagle Farm aerodrome, Amy allowed herself to be escorted by Stanley Bird towards the dais so that she could deliver her speech. She was dazed and doubtless suffering from delayed shock, and yet in spite of this she wouldn't let Brisbane down. It further enhanced her reputation in the eyes of many Australians who saw her as a girl of considerable courage who didn't quit, no matter what adversity chose to throw at her. Amy's outward show of courage was only temporay, however – a mask of her true feelings. For sometime later inside Government House, having endured an exhausting procession through the city, standing in an open-top car flanked by a mounted police escort, she finally gave way to tears. A doctor was called in, and strict orders were issued that Amy was not to pilot herself again, and so for the trip down to Sydney an airliner from Australian National Airways was made available: and Amy was afforded the luxury of being a passenger. She spent five hectic days in Brisbane before setting off for Sydney. During this time she learnt she had been awarded a CBE in the King's Birthday Honours List so generating yet another round of newspaper interviews with her parents. Will was also deeply concerned by the contents of a recent interview that Jack Humphreys had done with the *Daily Express* in which he had stated just how proud he was of Amy in very gushing words, which Will felt could mislead the public about the true nature of their relationship. Having already urged caution on Amy about what she said and to whom, Will Johnson was concerned that anything untoward could seriously jeopardise the very lucrative deal with the *Daily Mail,* which not only offered Amy the luxury of financial security, but for Will, always the businessman, the opportunity to recoup some of the money that he had invested in Amy's record-breaking bid. Any fears that Will had were to prove groundless in the case of Jack Humphreys but not so in general terms. For Amy was an obvious target for any exploitative fortune-hunter, with the charm to sweep her off her feet and thus inveigle themselves into her affections. The impact that a dashing young Scottish airman named James Alan Mollison would have on Amy, however subconsciously at their first meeting was nonetheless considerable. Mollison was co-piloting the *Southern Sun* Australian National Airways Avro three-engine plane that was taking her down to Sydney. Mollison was a flirt and an inveterate skirt-chaser with chiselled dark looks, blue eyes, and a devilish charm. Women simply found him irresistible, and Mollison basked unashamedly in his reputation. In the art of seduction he *always* demanded to be taken seriously and Amy caught his eye very quickly, so much so that he had elicited the promise of at least two dances with her at a formal ball to be held in her honour.

Whether this was simply a casual throwaway gesture on Amy's part isn't clear; however Mollison for his part had taken her invitation to dance seriously and duly turned up at the venue in Sydney, looking very dashing in his formal evening attire. Amy had been monopolised for most of the evening by Sir Phillip Game, the former Air Vice-Marshal and Governor-General of New South Wales whose ego was considerably flattered by having the most famous aviatrix in the world on his arm. He escorted her around, introducing her to the elite of Sydney society, whose status in the city afforded them the opportunity to meet the girl who had flown from the heart of the Empire.

Sir Phillip also danced with Amy and when the young co-pilot came to take Amy's hand it was Sir Phillip who gave him the brush-off, intercepting Mollison as he approached her. Sir Phillip said that Miss Johnson was extremely tired and would not be dancing again. Mollison was seething. Rejection by a woman was not something that he'd experienced very often, but he was wise enough not to tangle with a man of Sir Phillip's status. Amy merely shrugged her apologies, no doubt mindful of the fact that Castrol Oils would view the night's event as work and so expect her to be on her best behaviour.

While Mollison was unsure of the impact that he had made on Amy, he was in no doubt of the effect she'd had on him, but he was ruefully aware that on this occasion he'd been skilfully outmanoeuvred by an older man, and so the opportunity of spending some quality time with Amy had been cruelly snatched away.

Although Mollison may have appeared to Amy then as just another ardent admirer looking for his main chance, she was sufficiently intrigued by what she had seen to remember him. Even as a grown woman, she *still* yearned for the romantic 'fairytale' ending that she had read about in books as a child and while James Allan Mollison was cast more in the mould of a devilish 'black knight' than a dashing young prince, the impact on her feelings when it hit her some two years later, was as absolutely compelling as it had been, when as a 'gauche' teenager she'd first met her 'fair-haired' Swiss.

Although perilously close to an exhaustion-induced mental breakdown, Amy continued with a tour, which was yet to take her to Melbourne, Adelaide and Perth and trying her best to meet the demands Castrol Oils made on her to be constantly on show for her public, who turned out in their thousands to witness her arrival wherever she went in Australia. Sydney was a particular high point on her tour, and although closely shepherded by a Castrol Oils or a Shell representative by her side; Amy occasionally felt the urge to break free of her minders, and be herself.

Then along came *Smith's Weekly,* a Sydney-based tabloid which had pointed out somewhat scathingly that Amy appeared very keen to accept gifts, and thus she had her first taste of an adverse press. *Smith's Weekly* subsequently went for her in a particularly harsh manner. It could have simply been due to the fact that with so many column inches to fill the press were struggling to find new material to write about the Lone Girl Flyer. While Castrol Oils were determined to maximise every ounce of publicity they could get from having Amy as their prize it was inevitable that there would be some kind of media backlash, with Amy as the unsuspecting victim.

Smith's Weekly printed a 'spoof' letter from Amy to her parents satirising the true extent of her avarice which they sought to exploit further with headlines describing Amy as the 'Gimme, Gimme Girl' and the 'Air Digger of the Skyway' Amy was hurt both by the general tone of these comments, and the vitriolic intent that appeared to be behind some of them. But what angered her most was the inaccuracy in most of what they were saying. She would later give an interview to *The Truth,* which itself attempted to defend Amy by stating that their rival *Smith's Weekly* had blatantly misrepresented the details of the money that Amy had earned thus far just to fit the story they wanted to run regardless of how hurtful and potentially damaging their comments were.

Amy also felt justified in feeling let down by Castrol Oils whom she felt had done as little to insulate her from criticism in the press as they had in protecting her from the overzealous members of the public who rushed forward from a large uncontrolled crowd just to get a piece of her.

Whatever had motivated *Smith's Weekly* into going for Amy in such an obvious way it backfired on them quite spectacularly. Not only did Amy retain the support and the affection of the Australian public, she also enjoyed having other publications rushing to print articles defending her and damning *Smith's Weekly*, which one rival newspaper dismissed as being no better than a 'rag'.

Amy went on with her tour, visibly flagging with fatigue; she was often close to tears and in Perth, Castrol's representative Bill Brasch had arranged a publicity boost for a local aero club, which involved Amy flying a German-manufactured aircraft which she refused to do. While Amy lay face down on her bed sobbing, Brasch laid into her, yelling furiously, *"You're our servant. We bought you body and soul and you'll bloody well do as you're told."*

Whether the savvy but avuncular Lord Wakefield would have approved of such harsh behaviour by one of his agents is somewhat doubtful, but Ian Grabowsky who, having overheard most of what Brasch said, had already impressed Amy once with his kindness, readily stepped into the role of her 'Sir Galahad' warning Brasch never to shout at or abuse Amy again. Brasch told him to get out, reminding Amy that Castrol Oils claimed the rights over her until she left Australia.

In Perth another large crowd turned out to greet Amy, and so be a part of history. This was the last leg of the Australian tour, and for Amy, the end could not have come quickly enough. The finale though was not without controversy. When a man rushed from the crowd to give Amy a kiss, she slapped him hard across the face: an overreaction perhaps to an innocent and well meant gesture, which provoked gasps of shock and subdued laughter among the assembled crowd. Clearly embarrassed, the man disappeared. Amy however, was unrepentant. The incident and her unequivocal response to it served as a reminder that she was not the public's property and she wouldn't be treated as such. She was presented with a pair of boxing gloves for her actions in Perth, and she accepted the gesture and the humour behind it with good grace. There was however, a tiredness around her eyes, which told its own story.

On 7th July Amy stood on the dockside at Fremantle, posing for the last press photograph ever taken of her on Australian soil. Then she boarded the P & O liner SS *Naldera* and began the long journey home.

AMY ARRIVING AT BRISBANE

Photo © Daily Herald Archive/National Media Museum/

Science & Society Picture Library

10

CELEBRITY STATUS AT HOME & ABROAD

Long and lazy days aboard the SS *Naldera* restored Amy to the very best of health. She was able to stretch out her weary limbs with leisurely walks along the deck, while the paleness of her complexion and the shadows behind her eyes vanished as the sun and sea breeze gave her a healthy tanned look. She was given a suite of rooms on the P & O liner on the strict orders of Lord Inchcape, who, like other entrepreneurs, saw the commercial value in exploiting Amy Johnson's patronage to its fullest effect. Her star was in its ascendancy. Her celebrity status back home had reached a height that Will and 'Ciss' found alarming – as Amy would soon discover for herself.

Amy realised there must be a sizeable media frenzy gathering momentum around her when she learnt that aviation journalist Charles Dixon was rushing out a book about female aviators in general, which had her and her achievements at its heart. Its title, *Lone Girl Flyer*, was a direct reference to her. Amy had sought out her former employers with the idea of getting the publication of Dixon's book stopped. Crockers advised her that unless the book proved to be deliberately defamatory, there was little she could do to stop it, adding that litigation of that kind was not only prohibitively expensive but also invariably counter-productive.

Amy was also afforded the opportunity to reflect on the apparent estrangement from within her circle of admirers and of her mechanic and mentor at Stag Lane, Jack Humphreys, who had been gently pushed aside, largely due to her father's annoyance at Jack's indiscretion in an interview he had given to the *Daily Express*. Although she had been somewhat puzzled herself by what Jack had said as no plans, however vague, had ever been discussed for them to do a flight together and she couldn't help wondering at the less than benign influence of Jimmy Martin.

Amy had learnt, to her considerable cost, that when the press smell a 'good' story the truth is rarely allowed to get in the way. So, she had some sympathy for Jack Humphreys in his embarrassment at this indiscretion, and for the position outside her circle he now found himself.

Amy was invited to a formal state luncheon hosted by the King of Egypt on her arrival in Cairo. It was an extraordinary honour. This was the point at which the long and luxurious sea voyage ended. She had plenty of time to see the sights of historical significance in the ancient land of the pharaohs, to climb the Pyramids and visit the tomb of the boy king, Tutankhamun. Ahead lay the homeward journey by air with Imperial Airways to Croydon, refuelling at Crete, Athens, Salonica, Budapest and, finally, Vienna.

In 1930, the August bank holiday fell on the first Monday of the month – not on the last as it does today. So, on Monday 4th August, a large crowd thronged the streets of the South London suburb and waited in hushed anticipation for Amy's return. Some had been there for hours, determined to secure their vantage point and to witness a moment in history. Such was the high regard in which Amy was held in the public's affection that her homecoming had assumed a grandeur and importance akin to a royal visit. Perhaps only a Hollywood film star of the magnitude of Charlie Chaplin could have anticipated the crowd that gathered in the beckoning dusk of a summer evening to welcome Amy Johnson home.

The official welcoming party included the Mayor of Croydon in his robes of office and gleaming mayoral chain, the Member of Parliament for Hull, as well as Sir Sefton Brancker and Lord Wakefield of Hythe – the latter two men without whose help Amy may never have reached Australia. Also within the roped-off VIP enclosure were Will and 'Ciss' Johnson, who were trying their best to look and feel as if they belonged in such esteemed company. They waited for their eldest daughter to emerge into the harsh glare of publicity and to confront the daunting reality that, in the eyes of her public, she was now a national heroine.

Her Gypsy Moth biplane, *Jason,* had been fully repaired after the crash at Brisbane and, with new paintwork, was now sat gleaming on the tarmac. Rumours soon began to circulate that Amy would not be landing that evening and the crowd started to become slightly restless. However, they had already waited many hours for her and they would, of course, wait some more. The wooden dais on which she would stand to make her speech had been erected in readiness. She had spent some time on her homeward journey drafting said speech, after a cable from Will had forewarned of the reception she was likely to receive.

The rumours that Amy's arrival would be delayed eventually proved groundless, as a tiny speck in the sky over South London began to grow larger as it came closer. Then, the Imperial Airways airliner carrying Amy was overhead and the crowd responded in joyous rapturous applause.

They cheered and waved with anything they had to hand. Newspapers, hats, handkerchiefs. This was *the* day a nation had waited for since that fateful Saturday in mid-May, when, with their fingers crossed, they had listened to their crystal wireless sets and hoped that the Lone Girl Flyer would be found alive and well.

As Amy emerged into the glare of photographers' flashbulbs, the decibel of cheers was deafening and the collective sense of joy, boundless. The Hinkler record may have eluded her, but she had still become the first woman to fly solo to Australia. If a solicitor's typist from Hull could aspire to reach such heights, then every chambermaid in the best hotels in London could do the same. Waitresses in Lyons' Corner Houses and sales girls in the retail emporiums on Oxford Street could also believe that their wildest dreams were not beyond them. Amy was just like them and because she gave every indication of remaining so, the public adored her all the more.

Amy made a short speech that was rich in patriotism, but it was said in a strangely contrived accent, which those who had known her for years barely recognised as belonging to her. With her typical modesty, she apologised to the crowd for having kept them waiting for so long, and expressed her gratitude to them all. Then, sat high on the back seat of a car, Amy did two laps of honour around the tarmac of Croydon Airport so the crowd could get a better view. As Amy waved, the cheering grew louder. In an age before television, when people heard the news through listening to the wireless or reading the daily newspaper, or on their weekly trips to the cinema, the response to Amy's exploits in the skies and the general air of ecstasy around her subsequent homecoming was all the more remarkable.

Amy was nervous and perplexed when she discovered that the large crowd that had come out to see her was not, in fact, confined to the parameters of Croydon Airport, where they had claimed every conceivable vantage point to gain the merest glimpse of her. There were people lining the streets, in some places several lines' deep, despite the ever-enveloping darkness and the faint drizzle. Many had stood for hours.

Progress was slow as the convoy of cars meandered its way

through Croydon, Streatham and Brixton, before crossing the Thames at Westminster Bridge and continuing onto her new temporary home amid the splendour of Grosvenor House Hotel on Park Lane. Amy was obliged to stand on the balcony of her seventh-floor suite and acknowledge yet another crowd of exuberant well-wishers, before they would even consider going home.

The *Daily Mail* engaged the services of William Courtenay, as a kind of 'Man Friday', to assist Amy in many aspects of her professional life. They had promised her several days to relax and adjust to her opulent new surroundings and the ever-growing demands on her time. If the newspaper were in any way astonished by the enormous public response to her homecoming, they also saw an opportunity to maximise her commercial value through vigorous promotion while the public fascination with her was so great. Consequently, she was obliged to attend a gala luncheon at the Savoy Hotel on Wednesday 6th August, at which she would be presented with a cheque for £10,000 and a gold cup.

The guest list numbered 300 people and included eminent figures from the worlds of literature and the arts, such as J. B. Priestley, film director Alfred Hitchock and musician Ivor Novello. Malcolm Campbell, the land speed record breaker, was also there, and from the sphere of aviation the Frenchman Louis Bleriot and Bert Hinkler, came to pay tribute. Her credentials as an aviatrix of real significance could no longer be denied.

Readers of the *Daily Sketch* and the *Sunday Graphic* had been invited to contribute to a shilling fund to buy Amy a new aeroplane. This would be presented to her at a ceremony in Hyde Park one Sunday afternoon, in an event that would be attended by almost a 100,000 people. The aircraft, a Gypsy Moth, was to be called *Jason III*. The gift was in addition to *Jason II*, the name that Amy had given to a Puss Moth that she had already received from De Havilland, who were being very ambitious and somewhat courageous by marketing their new monoplane in such austere times. Gifts galore were being foisted upon Amy wherever she went, or they were delivered to her suite at the Grosvenor House. What she couldn't find room for or make use of was sent north to her parents. Couturiers wanted Amy to wear their clothes; milliners wanted her to wear their hats. Throughout the rest of 1930, in the sphere of product endorsement, Amy Johnson was one of the most sought-after names in England.

William Courtenay had quickly made himself invaluable to Amy and his role in her life was constantly evolving. From handling publicity and acting as speechwriter, he also took to dealing with the ever-growing fan mail. He was particularly zealous when it came to the letters containing proposals of marriage and other kinds of nonsense. He caused her some anxiety, however, by arranging a very extensive tour around Britain, which she was obliged to do under the contract she had with the *Daily Mail*. Her first date was in her home city, Hull, on 11th August. Amy was keen to have a holiday and so a few days away were arranged for her in Wales with her sister, Mollie, for company.

For the rest of August, Amy's time was taken up with the nationwide tour, which Courtenay arranged to concentrate on seaside towns popular with holidaymakers. So she crisscrossed the country, taking in venues like Great Yarmouth on the Norfolk coast and Bournemouth on the Southwest, with thousands of people coming out in soaring temperatures just to see *her*. In Bournemouth, almost 25,000 turned out in Meyrick Park, where Amy was to open the Bournemouth Hospital fête and where, just five years before, she had spent the lazy days of her convalescence strolling with a young, impressionable and utterly devoted Bunty Eddison, with her head full of romantic dreams and plans for the future. She had a secure, well-paid job in advertising and the man she loved on her arm.

Now, she attracted the attentions of more potential suitors than she could realistically cope with, let alone actually want. And the job that she really craved and for which she had undertaken the flight to Australia as a pilot in commercial aviation, based on merit and ability, still eluded her. In many respects, the success of her Australia flight had pushed that ideal further away than ever.

The tour soon took its toll upon Amy's health and by the end of August, rumours were circulating that she was in a near-state of sheer mental exhaustion. There was a serious danger of her suffering a complete nervous breakdown if she didn't start taking things easier.

When these rumours reached Will, he wrote to Amy, urging her to be mindful of her health. Then, in a more business-like tone, he told her to be aware of her current commercial value and the possibilities it afforded her while her star was at its zenith, but to be aware of those who sought to exploit her. For this reason, he was keen for her to start work on her autobiography as soon as possible stressing the desire that he and 'Ciss' shared that Amy should come home to live with them again.

Concern over Amy's health grew, and when she arrived in Brighton on the last weekend in August, Lord Rothermere wired her. Like Will, he urged her to take greater care regarding her health, while acknowledging the goodwill that she had already engendered through her nationwide tour. Although as keen as ever to extract the maximum from the investment he had made in Amy, he was a businessman of considerable acumen. He was acutely aware of the adverse publicity his Associated Newspapers might suffer if a national heroine like Amy Johnson was to collapse and have to be admitted into hospital while on an exhaustive tour that his company had arranged.

Amy was now so keen to see an end to the tour that when her doctor ordered her to have complete rest, she grabbed the opportunity to have it cut short, with almost two months' engagements outstanding. Executives at the *Daily* Mail, however, were keen for the tour to continue and suggested that their pilot should fly *Jason*. Amy reacted angrily and wouldn't sanction it under any circumstances. So, the proposal was dropped. *Jason* would be gifted to the nation and arrangements were made for him to be retired to a permanent home in the aviation department of the Science Museum. He still hangs there today, isolated in his smallness, and remains a lasting testament to Amy's achievement of flying to the other side of the world in such a fragile aeroplane of metal, fabric and wood.

In mid-September, Amy was staying at Norton Priory in the Sussex countryside. The break gave her the opportunity to relax, read, catch up with her correspondence and, given that she'd now been released from the demands of her nationwide tour, to do some serious thinking and make some crucial decisions regarding her future. She was very aware of how much 'Ciss' wanted her to return to the family home, and although she had no wish to hurt her mother's feelings, Amy was not convinced that such a move was in her best interests. She needed a place of her own and London was where she wanted to be.

Her suite at the Grosvenor House hotel had always been seen by Amy as a temporary billet, as was the offer that she eventually accepted to stay at the home of Lady Wiegall, where two rooms were offered to her in their house on Hill Street, behind Berkley Square, in the heart of the capital's fashionable Mayfair.

Amy wrote to her father while she was at Norton Priory, telling him that her stay there was just what she'd needed. The demands of public attention was wearying and she was just glad to be out of the public gaze for awhile. Still, she acknowledged that there was a sense of restlessness about her that she couldn't quite explain, and she knew that her lust for adventure would not lie dormant for long, and that another record-breaking flight might be the only way to assuage it. She also acknowledged just how close she had been to suffering a complete nervous breakdown. Nevertheless, Will and her former employer, Vernon Wood, were corresponding regularly and both were keen to offer Amy the moral support she needed.

Amy's recovery, although steady in those autumn weeks of 1930, was dealt a severe blow early in October when the airship *R101* crashed outside Beauvais in France, killing almost all its passengers as it burst into flames. Among the dead were Lord Thomson, the Secretary of State for Air, and Sir Sefton Brancker, whose death left Amy numb with shock and a raw sense of grief that she hadn't experienced since the death of Irene. The loss of 'Brancks' to the world of aviation was incalculable, as he had always been one of its most enthusiastic supporters.

Amy finally found herself a place to live and she moved into 15 Vernon Court, a modern-looking *pied-a-terre* in the vicinity of Finchley Road, which suited her needs perfectly. She soon had the company of two pet dogs: Rough, a bull terrier, who had been residing at Park Avenue with Will, 'Ciss' and the girls, until his mistress found them a place to live, and then Rex, an exuberant Red Setter, completed her family unit. She adapted quickly to the task of setting up home spending hours deciding which of the many souvenirs and gifts she had received in Australia should now adorn her new home.

Amy was, however, still in a state of limbo. Shortly after the *R101* crash, while the public was still absorbing the enormity of the disaster, Charles Kingsford Smith broke the solo record to Australia, taking days off her time. While Amy's achievement as a flyer was boosted when the International League of Aviators named her their air woman of 1930, it was actual times that counted in the arena of the record-breaking pioneer flyer. Amy realised that her nineteen and a half days now paled somewhat considerably in comparison to Kingsford-Smith's achievement.

Amy spent the Christmas of 1930 at home in Hull. There was some residual tension in the Johnson household following

Amy's outburst earlier in October, when she had accepted the hospitality of Sir Archibald and Lady Wiegall and had decided to adopt a pseudonym, known only to a select few, to keep the ever-invasive press at bay. Will had rung there to speak to her on some matter and had inadvertently asked for Amy under her own name, thereby risking blowing her cover. Although Amy later apologised to them, admitting that it wasn't their fault, it demonstrated to her their lack of understanding for the situation she was in, and their ignorance of the burden that her celebrity status had become. Added to this was her own trepidation about how to broach the subject of the flight she had planned for early January 1931 – to Peking via Russia, following the Trans-Siberian railway. To contemplate such a flight in the depths of a Russian winter in an open-cockpit biplane seemed hazardous, to say the least, and Amy knew that she was likely to attract some detractors in the press, eager to voice their opinions at the sheer folly of attempting such a flight in the most hostile of conditions. Her mother especially, would be sick with worry.

Amy was keen not to spoil Christmas, especially for her younger siblings, but with a proposed start date of January 1st she knew that this would be her only opportunity to tell her parents of her plans and so spare them the humiliating ignominy of finding out via the press. Will and 'Ciss' were numbed with shock. Recalling the twenty-four hours of sheer hell they had gone through when Amy was missing and feared dead at Atamboea, they couldn't understand why she had to put them through such an ordeal again by taking such an enormous risk. Yet, mindful of recent events, they were reluctant to alienate their daughter further by voicing their opposition to her plans. 'Ciss' took to hoping and praying that extreme weather and diplomatic procrastination in issuing permits to fly over Russia would curtail Amy's plans and keep her safe.

As Amy had predicted, once news of her flight plans entered the public domain, the sense of dismay at the enormity of what she was contemplating unleashed a barrage of criticism. Newspapers that had been largely supportive of her achievement in reaching Australia now expressed their anger. The *Daily Express* said that she was being 'neurotic'. Yet, despite all the invective, Amy's mind was made up. She left her two dogs in the care of Jack Humphreys and set off in *Jason III* determined to reach Berlin in one hop, before heading out across Poland and Russia against what threatened to be sub-zero winds. The Lone Girl Flyer was off again, and she would need all of her legendary 'pluck' and a very large slice of good fortune.

Adverse weather, the enemy that had shattered her hopes of setting a new record to Australia, was set against her once more. Her plans were quickly going awry. She was forced to make two unscheduled stops before she'd even reached Berlin, one overnight in Belgium and another at Cologne. She arrived at Templelhof aerodrome to be greeted by a small party, which included the wife of the British Ambassador, who offered Amy hospitality at the embassy for the night.

Still undaunted by the conditions, she was determined to set course for Warsaw the following morning, but the driving wind continued to dog her progress and with it came an ever-descending blanket of fog. *Jason III* was forced down by engine trouble, and when Amy landed in a field, she was distraught to discover that she had been forced some sixty miles off course. When informed that her aircraft couldn't be repaired in time for her to continue her flight, she took the news with as much good grace as she could muster.

Leaving *Jason III* in the care of her Polish hosts, she resolved to continue her journey to Moscow by train. Six days in and her plans to reach Peking via the Trans-Siberian route had been abandoned, as many in the British press would have readily predicted. Amy had lost a lot of respect, as well as the goodwill she had earned on Fleet Street.

The Russians, however, welcomed her warmly. Having

England's favourite aviatrix visit Moscow was a political coup they were eager to exploit. Whether Amy was yet ready or even willing to confront the reality, she had committed the ultimate sin of hubris. On her return, she would have to start repairing her image and another successful flight was the only way she could confirm her place in aviation circles. For Amy, the call of the Orient was proving irresistible and plans were forming in her mind for the next long-distance flight to the Land of the Rising Sun.

AMY IN "GIRL FLYER" STUDIO PORTRAIT

ss·A·Johnson
Arrived DARWIN from ENGLAND 24·5·30

JASON AT HENDON AERODROME

Photo © East Riding of Yorkshire Council

11

TO THE LAND OF THE RISING SUN

Amy returned home from Moscow chastened by her experience, and acutely aware that the press who had been critical of her plans from the outset could now barely resist the temptation to say, "We told you so". Amy still had much to be thankful for. Although her plans to reach Peking had been aborted due to engine trouble on *Jason III* and had forced her to land in a Polish potato field, the technical problem could instead have occurred in sub-zero temperatures over the frozen wastes of Siberia – in conditions in which she might easily have perished before a rescue mission could be attempted.

Amy accepted the invitation to join the Glass family on a skiing holiday in Switzerland, but still she was restless and ill at ease. The inability to settle down to anything like normal life had dogged her since returning from Australia. It worried her parents greatly, and it was having an impact on her health. So, the flight to Tokyo gave Amy something positive to focus on. She was also grateful for the moral support she received from Jack Humphreys. Her old mentor from the hangers at Stag Lane was back in the fold once more, despite Jimmy Martin's attempts to keep him out in the cold. Jack had endured his 'exile' with good grace, keeping his distance since the unfortunate misunderstanding over his interview with the *Daily Express*. He had convinced Will and 'Ciss' that he absolutely had no romantic feelings for Amy.

As a married man, it would have been entirely inappropriate. He had also made it clear that he'd never intentionally exploit her achievements for personal gain. Amy had been dismissive of Jimmy Martin's efforts to muddy the waters between her parents and Jack. In a letter to Will, she described Jimmy Martin as the "biggest mischief maker" she knew.

In February 1931, Amy was invited to lunch at the home of George Bernard Shaw and his wife, Charlotte. The occasion was recorded for posterity with a now famous photograph of Shaw, Amy and two other guests. One was Lady Nancy Astor, the American born MP who, nearly three years before, had chaperoned her compatriot Amelia Earhart on her tour around London. Lady Astor was a keen supporter of aviation in general. As the first woman ever to take her seat in the House of Commons, she was renowned as being a vigorous supporter of women and of applauding their achievements in any field. The quartet was completed by Charlie Chaplin, the most recognised face in silent cinema, who was paying a rare visit to his homeland. Amy confessed to having been an ardent fan of Shaw's work since her adolescence, so the invitation to lunch had offered her the opportunity to meet one of her literary heroes. In a letter to Will, however, she confessed to being less enthusiastic about meeting Chaplin.

Jack Humphreys eagerly accepted Amy's offer to accompany her as mechanic and co-pilot on her flight to Japan. She eventually decided to take *Jason II*, the Puss Moth that De Havilland had given her. They planned to set off from Lympne on the Kent coast at 12.30am on 28th July. Frequent refuelling stops meant that while Jack took the controls, Amy could grab a quick nap. It was a luxury she hadn't enjoyed on the Australia flight. It also enabled them to cover greater distances in a day. Although they had three stops for refuelling before Moscow, they still managed to cover 1,700 miles and set a new record of reaching the Russian capital in a single day.

Although Amy was quietly satisfied with this new record, she refused to be overawed by it. After the public relations disaster of her aborted flight to Peking, Amy believed she still had something to prove to her critics back home and so she was determined to remain focussed on having a serious go at the England to Japan record. They followed the route of the Trans-Siberian railway and were constantly referring to their maps in fear that they might lose sight of it, in its meanderings through the dense forest. They had overnight stops at strange and fascinating places with hard-to-pronounce names like Kazan, Kurgan, Taijan and Irkutsk.

The Empire of Japan had been flexing its muscles militarily. This had become a disturbing complication and one that Amy could have done without while she had been planning the flight. Getting entangled, however unwittingly, in a complex and tense political situation was always concerning and the Japanese had made their ambitions regarding Manchuria abundantly clear by invading the territory, belonging to China, before they'd set off from Lympne. Amy and Jack Humphreys still managed to enjoy the hospitality of the British Consul at Mukden, the Manchurian capital where they stayed overnight, and then they flew onto Seoul, Hiroshima and Yokahama. On 6th August 1931, they caught the first glimpse of Mount Fuji as they began their descent into the Japanese capital.

They set a new record of ten days travelling from England to Japan, but if Amy thought achieving this would put her back on the front pages, she was wrong. She was the unfortunate victim of bad timing. Wiley Post, the American pilot, had circumnavigated the globe in just over eight days, while the dashing Scotsman, Jim Mollison, who had flown Amy to a gala ball in Sydney just over a year ago, had recently secured the backing of Lord Wakefield for an audacious bid at breaking the Australia-England record.

Mollison landed his Gypsy Moth on the beach at Pevensey to great acclaim, after just over eight days in the air – an achievement that saw him take two days off the record recently set by C. A. Scott. Mollison arrived on the same day that Amy and Jack arrived in Tokyo. Piqued though she was not to garner more publicity for setting a new record herself, Amy knew that in the golden era of the pioneer flyer, there were a great many records set and broken as raw ambition and competitive spirit motivated pilots to slice minutes, hours and sometimes days off an existing record and knock their rival off the front page.

The Japanese were fascinated with Amy and she had already gained some notoriety with them by virtue of being the first woman to fly solo to Australia. She was the subject of considerable curiosity to her hosts and they were courteous, generous and very obliging in their hospitality towards her. Amy and Jack knew they had to be astute in their response to this outward show of goodwill. Their tact and diplomacy was vital. Japan had raised the international temperature by sending its troops into Manchuria and every diplomatic effort had been deployed to ease the tension between Tokyo and Peking. Amy couldn't afford to be seen as a pawn to Japanese Imperialism, especially as she had been heavily criticised, particularly in the Conservative press, for having allowed herself to be photographed with Lenin's widow, Krupskaya, and other top Soviet officials, on her trip to Moscow earlier that year.

Amy and Jack had planned a week-long stay in Tokyo, and while her achievements had received scant attention in the press, she had set a new record to Japan, which she believed merited greater recognition. The Rotary Club of Tokyo invited her to a formal lunch and to make a speech afterwards. There were other engagements to attend at the British Embassy and a ball organised by the Imperial Aviation Society, at which she was guest of honour.

Jack Humphreys kept himself at a discreet distance, happy for Amy to bask in the limelight while he acted as a kind of consort. He had accompanied her as co-pilot and mechanic, ensuring the Puss Moth was in prime condition for their return flight was his priority. He also accompanied Amy on a visit to a Japanese hospital to see Francis Chichester, who had suffered an horrific crash. Chichester would later achieve even greater notoriety as an international round-the-world yachtsman; and it was for his prowess on the high seas that he was awarded a Knighthood, after his many achievements and occasional mishaps in the air.

While Jack managed to quash any lingering rumours about him and Amy being involved romantically, there were inevitably other potential suitors, who the press loved to link her with. One such man was Peter Q Reiss, a dashing sandy-haired amateur pilot, who was attached to the Leicestershire Flying Club at Ratcliffe. He first met Amy at a weekend social event. Reiss was a Lloyds underwriter and an accomplished squash player to championship level. He was good-looking and wore the fact with a touch of urbanity. He was, however, initially quite earnest in his adoration of Amy and, according to her friends, she found his adoration too strong for comfort. However, they remained friends and in later years he became an valuable source of moral support.

Amy and Jack returned home to find that London society was agog at the flying exploits of Jim Mollison. How Amy felt when she learnt that Mollison's achievement in setting a new record from Australia to England was responsible for sweeping her off the front pages is unclear, but there was no escaping the easygoing charm of the dashing Scotsman. The young debutantes and their mothers were utterly captivated by him. Mollison was more than willing to devote his time in satisfying their curious young minds, especially when someone else was paying.

Amy had a nationwide tour to get on with, which would consume most of her time and energy. Will Johnson had made his feelings regarding the lack of publicity around the Japan flight clear, stating that he could only hope the upcoming tour would earn her the recognition that setting a new record deserved. Will had vexed his frustration in a letter to William Courtenay. He had also made several important decisions about his future, edging nearer to full retirement from 'AJK' and severing his links with Hull, moving from Park Avenue to a bungalow in Bridlington. Although Betty Johnson, his youngest child, was twelve years old and still his responsibility, he had acknowledged that the time had now come for him to adopt a more relaxing way of life.

Jack was accompanying Amy on the tour, which would be conducted on a more relaxed basis than the aborted nationwide lecture tour that the *Daily Mail* had arranged for her the previous year, whose proprietor, Lord Rothermere, had felt obliged to release her from on health grounds. Amy was still plagued by ill health and a general sense of fatigue, but she spent Christmas 1931 with her parents in their new Bridlington home. Her sister, Mollie, had recently become engaged to Trevor Jones, a Welshman who had impressed his future father-in-law by securing a well-paid job as the Deputy Town Clerk for Blackpool Council.

Amy suffered from excruciating period pains and each month her menstrual cycle would have her bent double in agony. She was also prone to severe black moods, which had a dramatic impact on those around her. Consequently, she felt unable to function properly. She was struck down again early in 1932 after an engagement in Bolton. So that February, with the support of her family and physician, Amy made a life-changing decision and admitted herself to a hospital in London for an appendectomy – as the press had chosen to report it. In truth, Amy was to undergo a hysterectomy. It was a painful and very traumatic decision, both emotionally and physically, as it ended her hopes of experiencing the joy of motherhood.

Amy found convalescing after her operation boring and tiresome. English weather in February was hardly conducive to a quick recovery. She decided that rather than continue to endure the dreariness of an Essex-based convalescent home in winter, she would take a short trip away to put some colour back into her cheeks and to allay her growing sense of frustration. Her admission to hospital had decimated her tour schedule and so she had the luxury of time on her hands. With this in mind, she boarded the liner *Winchester Castle*, which was bound for Madeira.

When the *Winchester Castle* arrived at Madeira, the normally balmy Portuguese island was suffering its worst spell of weather in years. Amy's mood was not sweetened when she arrived to find her hotel was largely occupied by stiff middle-aged matrons, with whom she had virtually nothing in common. So, she ordered that her luggage be returned to port and re-embarked the *Winchester Castle* to continue the journey to South Africa.

Amy spent the rest of the onward journey to Durban writing letters to her parents and sending cables of instruction to Jack Humphreys. She had decided some time ago to have a tilt at the record for circumnavigating the globe, which had recently been set by Wiley Post and his flying partner Harold Gatty. They had set the new record shortly before Amy had broken the record to Japan. On her return, she had set her plan in motion by contacting the Vickers aircraft company with this in mind. Amy spent the rest of the voyage soaking up the sun, as yet unaware that Jim Mollison was having a tilt at the solo record to the Cape. The ubiquitous dashing Scot seemed to be just about everywhere. Now, he was to make another big dent in Amy's carefully laid plans.

In truth, the press were more interested in the growing rumours of a romantic entanglement involving Jim Mollison to make anything of the easy rapport that seemed to exist between the two flyers. While Amy helped Jim Mollison to sort through the mountain of cables congratulating him on his heroic flight to Cape Town, the press were more interested in the ongoing rumours surrounding him and Diana Wellesley, the eighteen-year-old great granddaughter of the Duke of Wellington. A possible romance between a handsome daring flyer and self-confessed ladies' man, and an attractive blue-blooded member of the aristocracy was meat and drink to a gossip-hungry press.

Individually, Amy Johnson and Jim Mollison had been a fascination for Fleet Street hacks ever since they had forced their way to the front of public attention. Now, even the slightest snippet of news about them, their record-breaking attempts, or the occasional brushes with danger, filled column inches and sold newspapers. When brought together – for a few years, at least – they would be an unbeatable team.

AMY ON BOARD SHIP FROM CAPE TOWN

Photo © East Riding of Yorkshire Council

12

AMY & JIM

"A born flyer if ever there was one" is how a fresh-faced and youthful looking James Allan Mollison was described by one of his instructors at the RAF Flying School at Duxford near Cambridge.

The praise that was heaped upon the young novice was qualified, however, with the observation that he was a bit headstrong and that he would need watching. Regardless, the young Scot's natural flair for flying was obvious to the trainer's eye at a very early stage.

It also became clear that Mollison had a less-than-rigid respect for the rules, and one incident, in which he foolishly tried to test their elasticity, could have potentially been very damaging to his future with the RAF. During training exercises one December afternoon in 1923, when the fog that had hung stubbornly over the Cambridgeshire fenlands since early morning finally began to disperse, Mollison was sent up in an Avro training aircraft and told to remain strictly within the aerodrome's airspace. Mollison had already decided that a short trip to see a casual girlfriend who was studying at the nearby Girton College was too much of a temptation to resist, and one that he believed he could manage without any of the instructors noticing he had gone missing.

He flew over the Girton campus, imagining somewhat egotistically that the students were waving up at him from the grassy quadrangle below. He did a couple of showy turns and then headed back towards Duxford, hoping that nobody would have noticed his absence – except that he came down in a field some ten miles off the aerodrome perimeter, upside down in the cockpit. Miraculously, he didn't suffer any serious injuries. His Avro aircraft, however, did not fare so well and what he had considered to be just a harmless detour would now become a serious disciplinary matter, that could even lead to his dismissal from the RAF. He called the base, asked for the commanding officer and steeled himself for the very daunting task of telling them exactly what had happened.

When Mollison and the damaged Avro biplane arrived back at RAF Duxford, he was told to report immediately to the Station Commander, a WWI veteran and holder of the DSO, who demanded a full explanation of why a trainee flying officer had disobeyed a direct order. Eventually, Mollison was dismissed from the Wing Commander's office and, that evening, was subjected to some considerable leg-pulling in the station mess as news of his jaunt became common knowledge. He knew that he could easily have been on his way home to Scotland, drummed out, humiliated, and forced to explain his actions to his family and incur their inevitable wrath, but he'd had a lucky escape.

James Allan Mollison – or Jim, as he was known throughout his life – was born on April 19th 1905 in Pollokshields, Lanark, near Glasgow, to Hector Alexander Mollison and Thomasina Addie. The marriage was a turbulent one and Jim was their only child. Thomasina or 'Tommy', as she was known to her family and friends, had been raised in a strict Presbyterian household and thus she took a very dim view of her husband's weakness for alcohol. This, along with the tendency towards violence that alcohol frequently provoked in him, was the main catalyst in the failure of their marriage.

Thomasina tolerated her husband's unpredictable behaviour for as long as she could manage before taking the drastic step of terminating the marriage. However, she still wanted out with her dignity intact. She wanted to divorce Hector on the grounds of his desertion, and it was with her family's money that his freedom was won. So, with a sum settled upon him and passage bound for Australia aboard one of his brother-in-law's ships, Hector Alexander Mollison vanished from his son's life.

Jim Mollison took life at RAF Duxford in his stride. The practical flying exercises came easily to him, and overall his piloting instincts were sound. The disciplinary aspects, however, were a little harder for him to take. He did apply himself to the examinations that were essential to him passing out as a flying officer and earning his wings, but he soon learnt that coming in the top category had its drawbacks. It generally meant a home posting and he had itchy feet. He wanted a taste of the exotic, which meant a posting overseas.

He became one of RAF Duxford's youngest pilots and was posted to RAF Kenley in Surrey, whose proximity to the bright lights of the West End suited him better. Then, towards the end of 1924, he was posted to a base near Grantham in Lincolnshire, prior to the overseas posting he had so craved. He had come a very long way in less than a year, when he had thought he could flirt with authority and bend the rules.

Thomasina accompanied Jim, who had been staying with her in Scotland while on leave, to Southampton, where the troop ship *Assaye* for Bombay awaited him. For the boy who had once watched countless ships sailing down the Clyde, out of home waters and heading for the far-flung corners of the globe, the days of dreaming were over. He was now a young man, and although not yet twenty, he was about to enjoy his first taste of foreign adventure.

As the days and weeks rolled by, Jim and his RAF colleagues, along with the 300 or so army troops that were with them aboard the P & O ship *Assaye*, tried their best to relax as they made their way to Bombay. An acute sense of boredom was a natural reaction for a man as active as Jim, but the voyage afforded him the opportunity to prepare mentally for the potential dangers that he faced on his first overseas posting.

Jim Mollison always believed in living for the moment and money invariably burnt a hole in his pocket, thus he enjoyed spending it as quickly as he could earn it. He was determined to soak up all the pleasures that an exotic, teeming and vibrant city like Bombay had to offer, and his plan of decadent living began with booking himself into the Taj Mahal Hotel, which enjoyed a panoramic view of the city. He was waiting further orders regarding his posting and so the first ten days in Bombay were his own. He was determined to cram in all the sightseeing he could manage before moving onto Karachi.

The North West Frontier Province in which Karachi was then situated had been experiencing some fierce and determined resistance to the benign influence of British Empire rule. After the partition of India in 1947, it became part of the modern state of Pakistan. Because of its proximity to the Afghan border, it had been the scene of many skirmishes. It represented the most serious challenge to ongoing British rule in India, which the warrior tribesman who dominated the arid mountainous border area were determined to bring to an end.

Jim served for almost three months in Waziristan, flying Bristol F2B bomber aircraft as part of Bomber Squadron 20 against the formidable Mahsud tribesman. Their determined efforts to regain control of Waziristan had forced the British to nominally recognise Afghan independence four years earlier, and having now established their reputation, they were determined to consolidate and further undermine the British influence in India once and for all. The conflict involved almost sixty days of continuous heavy fighting, in which the RAF crews flew over 2,000 hours of sorties before the Mahsud agreed to come to the negotiating table. If Jim Mollison had wanted a baptism of fire on his first overseas tour of duty, then his wish had been granted. For the politicians and the civil service mandarins, there was one sobering reality they could not afford to ignore, and the moral behind it was unequivocal. The benign acceptance of the British in India would not be tolerated in perpetuity. Whatever firepower they had to use to quell the enemy, resistance to them was now stronger, better organised and constant.

His tour of duty in India lasted twenty-one months. After the frenetic action that he had experienced during the conflict in Waziristan, he was stationed at Lahore. For Jim, this was a period of tedious boredom, endured in unbearable heat. Luckily, it was mercifully punctuated with short bouts of leave. His posting back to England came at his request, citing private reasons. It was granted some months before his tour of duty had officially expired, and so in September 1926 Flying Officer Mollison boarded ship and headed home.

Jim Mollison served the last seventeen months of his five-year Short Service Commission in the RAF at bases dotted around the country. His first posting on returning to England was at the Electrical & Wireless School, near Winchester Dorset. He arrived swarthy and a little battle-weary after his experiences in the NWFP, and he had lost the boyish looks that he had possessed on entering Duxford. His new role as Secretary to the CO meant he was left with mundane administrative tasks, which quickly induced boredom, and thus he quickly adapted to the subtle art of delegation. He stuck it out at Flowerdown for as long as he could before putting in for a transfer, which was accepted and took him to the Central Flying School at RAF Wittering in Lincolnshire. With its numerous RAF bases dotted around the county, it earned the nickname 'bomber county' during the

Second World War. It was familiar territory to Mollison, as he had briefly served at RAF Spittlegate near Grantham. Now, he was in his element as a pupil-instructor. The mind-numbing tedium of his three-month stint as a pen-pushing desk jockey were over and he was back to where he believed he belonged, in the cockpit, passing his considerable piloting skills onto a new crop of raw recruits.

His three-month course at CFS garnered him an almost 80 per cent pass mark, and in April 1927 he was beginning to consider his future outside the RAF. Flying would always be important, and it came naturally to him, but he knew he'd never miss the rigid formality and the rules that came with service life. His last posting was at Number 5 Flying School, RAF Sealand near Chester. He spent almost eleven months there and as time ticked on, he was itching to go and put his piloting skills to the test in a competitive environment. There was also the gratuity of £350.00 that he would receive on his release and it was burning a hole in his pocket before it had even got there.

Jim Mollison had been a pilot with Australian National Airways for just a few months in mid-May 1930, yet his travel horizons and overall life experiences were considerably broader than those of the exhausted girl-flyer who'd crashed her beloved Gypsy Moth biplane, *Jason*, into a perimeter fence at Brisbane's Eagle Farm aerodrome, and was now being flown as a VIP passenger to a gala ball that was to be held in her honour.

He had seen action on India's North-West Frontier and since leaving the RAF had lived a nomadic, but somewhat decadent, life on the French Riviera. He had also enjoyed the exotic delights of Tahiti and then, having whittled away most of his £350.00 gratuity, had been forced to accept menial work on the beaches around Sydney. This contrasted wildly to the life of the solicitor's typist from Hull, who had tried her hand at various careers before finding her niche in the hangers at Stag Lane. Her only experience of overseas travel had been a brief holiday in the

Swiss homeland of an ex-boyfriend until, that is, she made the momentous decision to fly to the other side of the world.

Although Jim found himself attracted to the diffident, moody Amy and knew instinctively that he liked her, he couldn't help reflecting somewhat ruefully on the enormous irony of their situation. Although Amy and Jim lived in London during the latter part of 1931, their respective worlds didn't dovetail until they met again in Cape Town in the spring of 1932. Despite the less than auspicious nature of his arrival in Cape Town by landing his Puss Moth upside down in 5 feet of seawater, he had broken the record from England – via the trans-Sahara route – by a staggering fifteen hours. Amy was still feeling emotionally fragile after her hysterectomy operation, but in the relaxing atmosphere of their surroundings, they reaffirmed their friendship and discovered, on a subliminal level, a connection that went much, much deeper.

Amy was met at Southampton on May 5th by 'Ciss' and Jack Humphries, who quickly saw a difference in her demeanour. She was tanned from the long hours spent lying in the sun, recuperating at the clifftop home of a South African millionaire that enjoyed panoramic views of the Indian Ocean, but they couldn't work out the cause of her upbeat mood. Four days later, however, when they returned from an air display in Brussels to find Jim Mollison waiting for Amy at Heston aerodrome, Jack believed he had his answer. Amy was in love.

She accepted a lunch date with Jim at the fashionable London restaurant Quaglino's the following day, feeling relaxed and in a much better state of mind than she had enjoyed for awhile. She had taken time in South Africa and on the sea journey home to think about the future and to weigh up her options. Her flight to Japan the previous summer had set a new record, but it had failed to inspire the press. Now, she had been forced to accept that Mollison's successful attempt at the Cape record had helped to make him Fleet Street's latest flying 'star' while the continued rumours about his engagement to Diana Wellesley only served to boost his profile further. Jim was dazzled by the tanned and sophisticated Amy, who looked considerably more feminine than he'd remembered her being in Australia two years ago. Blonde and blue-eyed, she was exactly the type of woman to catch his eye, and her flying exploits lent her a certain cachet that some of his other conquests lacked. She belonged in plush surroundings like Quaglino's, and, with her confidence brimming, the conversation flowed easily between them. Flirting had always come naturally to Mollison; it was an art that he never grew tired of practising. Even so, it was still quite a leap for Jim to propose marriage to Amy over their brandy liqueurs. Amy accepted his proposal and, their meal over, they made haste to the offices of *The Times* and announced their engagement to the public. The news of Amy and Jim's engagement caused a stir on Fleet Street. In 1932, Amy was the most famous woman in England. Even the Queen of Crime, Agatha Christie, had been sufficiently impressed by Amy's achievement in flying solo to Australia. Although Christie didn't name her directly, it was clearly Amy that her characters were referring to in the novel *Peril at End House*. Although her celebrity had dipped in recent months, even a snippet of news about her aroused the public's imagination. Only Gracie Fields could rival Amy, in terms of public adulation. The Rochdale-born songstress had endured a modest upbringing and yet she demonstrated the same pluck on stage that Amy showed in the air, becoming the highest paid entertainer of her day.

The engagement was still viewed with some scepticism by the press, however, as many hadn't seen it coming. Indeed, there had been no telltale signs of a romance between them. Fleet Street had been expecting a formal announcement of an engagement between Mollison and Diana Wellesley for some time, as the teenaged debutante had made no secret of her infatuation for the dare-devil flyer. She didn't care who knew it and these reckless public displays of affection had caused deep concern within her family, prompting her stepmother, Lady Clare Cowley, to summon Jim down to the family home, Seagray Manor, near Chippenham, to confront the problem of his on-off engagement to Diana head-on.

Lady Cowley told a very nervous Jim Mollison that marriage to anyone was non-negotiable until Diana had reached the age of twenty. She grilled him about his personal life and financial affairs, before telling him unequivocally that, in terms of breeding and social status, he simply didn't measure up to what the family would expect from a potential husband for Diana. Lady Cowley also made it clear that unlike her stepdaughter, she was not impressed by his record-breaking achievements in the air. Diana, who had been waiting anxiously in an adjoining room, was deeply disappointed, as the feelings on her part were genuine. She was besotted with Mollison, to the extent of suggesting an elopement, but Jim was too shrewd to succumb to such a romantic notion as eloping. He knew the implications, for himself, of such a drastic move.

Jim certainly didn't cover himself in glory when it came to telling Diana about his engagement to Amy, and choosing to take her to Quaglino's was a tactless move on his part. When Diana was approached by the press for her comments on the Mollison-Johnson engagement, she demonstrated the refinement and stiff upper lip typical of her breeding and class by wishing them well for the future. It is likely, however, that the engagement of Britain's best known flyers was received with relief and joy in the drawing room of Seagray Manor.

'Ciss' Johnson was told on the same evening by Amy herself and was instinctively concerned at the pace with which events were unfolding. They hadn't been aware of Jim Mollisons' romantic interest in Amy and they were always wary of potential gold diggers. Will met Jim the next day with Jack Humphreys, and was unimpressed by the man that his daughter had decided to marry. The meeting took place in the plush surroundings of the Grosvenor House Hotel. If Will believed that Amy was being used by the Scotsman, he certainly wasn't alone in his theory. She was far ahead of her fiancé in terms of celebrity value and many people didn't accept this as a genuine love-match. Amy quickly picked up on the scepticism, and a week after news of her engagement had appeared in the press, she wrote to her father, urging him to accept that she was very happy and doubted that she'd regret the decision.

The wedding took place at St George's Church, Hanover Square, on Friday 29th July 1932, just over eleven weeks after they had got engaged. There had been further speculation in the press about whether there would actually be a wedding, amid lingering suspicions held by some that it was all just an elaborate publicity stunt. Amy and Jim had been seen and photographed together as often as possible during those weeks, to prove their detractors wrong and to stem the flow of negative press reports that said they were getting engaged to enhance their public image. Towards the end of June, they took a holiday in the fashionable French resort of Juan-le-Pins. If Amy had harboured any serious doubts about the man that she had consented to marry, this would have been her chance to call it off. Jim demonstrated his usual weakness for alcohol, which, along with his habit for allowing money to burn a hole in his pocket, would have given her cause for concern.

Amy had been in touch with Jack Humphreys prior to leaving for France to inform him, her friend and mentor, that his services were no longer required. What Jack must have thought about this dramatic severing of links can only be imagined.

The decision that Amy took not to invite her parents to the wedding was a difficult one. She had followed Jim's lead in not inviting his mother, Thomasina Bullmore, with whom his relationship had always been constrained, but even in its carefully managed execution, Amy still contrived a public relations disaster, which caused Will and 'Ciss' Johnson unnecessary hurt and embarrassment.

Amy had cabled her parents on the eve of the ceremony – Thursday 28th July – to tell them that she was getting married at ten o'clock the following morning. Horrified that they had been left with so little time to prepare and doubtless eager to pin the blame for a deliberate snub squarely on the shoulders of Jim, they left Bridlington in the early hours of the morning, determined to reach London in time for the wedding. They arrived at St George's Church after the ceremony as the couple were signing the register. 'Ciss' was particularly hurt that she had missed seeing her eldest daughter get married, and although they went into the church, they chose to remain seated at the back, rather than join the invited guests. Amy had been given away by Kathleen Countess of Drogheda, thus robbing Will of the role that was rightfully his.

Jim had chosen Francis Shelmadine, the director of Civil Aviation, as his best man. Concealed at the back of the church, Amy's family suffered the further indignity of not being seen by Amy and Jim as they walked down the central aisle and out of the church.

The reception at the Grosvenor House Hotel was well under way before anyone mentioned to Amy that her parents had been recognised in church. She soon had people frantically ringing round the hotels that she knew her father used on his business trips to London, but she was too late. Her family had left for Bridlington almost immediately, deeply upset that their daughter could be so complicit in a slight against them and on such an important day in her life.

Amy was distraught; the revelation that her family had been in London and that she hadn't seen or spoken to them soured the day. The reception itself had been somewhat subdued. Jim was preoccupied with his forthcoming flight across the Atlantic, while Amy was far from acting the happy blushing bride. Her speech at the reception was short and succinct, while Jim had his usual air of nonchalance. They left London later that day in separate aeroplanes for a short honeymoon in Scotland, having tried to avert any press intrusion by telling them they were going to Le Touquet.

While staying at Kelburn Castle on the west coast of Scotland as guests of Lady Bowden, Amy and Jim climbed the battlements to take in the fantastic view and to get a glimpse of the Atlantic: the vast ocean that Jim would have to negotiate in his bid to make the first solo crossing in the east-west direction. 'The Water Jump', as it was known, represented the greatest challenge to the pioneer flyer: the longest period of flying without sight of land. It was a true test of a pilot's courage and powers of concentration. He would need enough fuel in the tanks of his Puss Moth *Heart's Content* to get him safely across the ocean, favourable weather conditions and, as with all long-distance flights, a large slice of good fortune.

After much deliberation, Jim chose Portmarnock Strand near Dublin as his take-off point, as it boasted a long stretch of smooth sandy beach. His Puss Moth had been fitted with extra fuel tanks for a flight of almost 3,200 miles to New York. Having thanked the crowd that had enthusiastically gathered to wave him off, Jim climbed into the cockpit of the *Heart's Content*.

For Amy, the waiting was tortuous. She had never been in the situation of having to endure the endless hours of a long-distance flight when the fate of someone you loved hung in the balance. Now, she had a sense of what her parents must have endured on her epic flight to Australia, after Will had stood on the tarmac at Croydon and watched until *Jason* was just a tiny speck in the sky. She had to hope now that Jim could pull it off, and then all the accolades and newspaper headlines that such an achievement would bring him would be richly deserved. For awhile, at least, she would stand aside and bask in his reflective glory.

Jim landed at New Brunswick after more than thirty hours

in the sky. He had thought that he had enough fuel to press on and land at Roosevelt Field, Long Island, without having to touch down at Harbour Grace, but he had been over optimistic. Nonetheless, he had done it. The fastest solo flight across the Atlantic in the east-west direction. Bill Courtenay could barely contain his excitement, when he told a strangely subdued Amy that Jim had achieved his ambition. Jim was ecstatic, and demonstrated his usual exuberance and willingness to accept the generosity of others in celebrating his achievement. He had a private meeting with Charles Lindbergh, and the movie star Douglas Fairbanks Junior also expressed a wish to meet him. Jim was having too good a time being idolised wherever he went, and for awhile he put all thoughts of returning home to the back of his mind.

Jim had always wanted to fly back home in the *Heart's Content* to see if he could achieve the double by setting another Atlantic record in the easterly direction, but his sheer nervous exhaustion and the unfavourable weather spoilt his plans. On the telephone and by cable, Amy urged him not to risk it, but Jim could not be persuaded to abandon his plan easily. A cable from their sponsor Lord Wakefield urged him to heed Amy's advice and Jim finally acceded to his demands.

A smiling and much-relieved Amy met her husband off the ship at Cherbourg. A reception had been arranged for him at the Grosvenor House, which his mother, Thomasina Bullmore, would attend along with Will and 'Ciss' in a gesture of conciliation towards their son-in-law. Amy was just glad to have Jim back safe. The thought of widowhood, after just one month of marriage, was something that she had *forced* herself to think about, but chose not to dwell on.

It was an habitual hazard for the long-distance flyer. Amy was aware of what the press expected of them. As Brand Mollison, they had to keep on setting and breaking records to stay on the front pages. Jim had done his job and now it was her turn. Being the first woman to do the 'Water Jump' alone was a record that Amy had coveted for awhile, but Amelia Earhart had claimed it for herself. So, Amy set herself a new target of smashing Jim's record to Cape Town, via the trans-Sahara route.

13

SOLO TO THE CAPE

Amy was very conscious of the need to make some kind of splash on her flight to the Cape. Her previous record-breaking flight to Japan in the summer of 1931 had failed to inspire the press, and with a plethora of record-breaking flyers now on the scene, there was little appetite on Fleet Street for paying out thousands of pounds just so their readers could have the privilege of reading how the latest stunt-flyer had shaved minutes off an existing record. The double record from London to Cape Town to London, in an overall record time, seemed to be her best bet of making an impact.

Amy acquired a new aeroplane for the trip, a De Havilland Puss Moth, and she broke with her usual tradition by naming it *Desert Cloud*. She left Bill Courtenay with the responsibility of negotiating the best price that he could get for exclusive rights to her story. It required all of his usual deftness, and being a pragmatist, Courtenay knew it would be difficult even with a client as famous as Amy. It meant that advertisement through personal endorsement became ever more important as a means to boosting her income. Ever since she'd returned from Australia, Amy had been eagerly sought after by anyone with a product to sell. Couturiers, such as Elsa Schiaparelli and Coco Chanel, would do anything to have her wearing their clothes. Any aeroplane or motor-car maker who could claim that Amy Johnson was flying or driving the latest model of their design could be certain that orders would flood in, and a magazine such as *Tatler* or *Vogue* with Amy on the front cover would justifiably expect a boost in circulation. Throughout the 1930s, the decade that she could rightfully claim as her own, Amy enjoyed a level of celebrity status similar to that which Diana, Princess of Wales, attained half a century later.

Amy was determined to leave nothing to chance and enrolled on a short course of blind night flying so that her navigational skills wouldn't let her down, and this time she had some experience of long-distance flying. The trans-Sahara route would be very unforgiving should she get lost, but it was the most direct route, which Jim had used on his record-setting bid back in March. It took 600 miles off the distance compared to the route favoured by Imperial Airways, from Croydon via Cairo, Khartoum, Kisumu and Johannesburg.

Amy set off from Lympne on the Kent coast shortly after 6.30am on Monday 14th November. The small crowd that had gathered to see her off was illuminated in the pre-dawn by a flood-lighting truck, and among them were her secretary and 'Man Friday', Bill Courtenay. Amy and Jim kissed, then climbed into their respective cockpits. Jim was flying the *Heart's Content* as far as France, before taking his leave. This was Amy's show, and however ambiguous Jim may have felt about the prospect of his eight-month-old record being broken, he was keen to be seen supporting his wife.

Amy planned to reach Oran in North Africa in one hop, but a suspect fuel gauge forced her down in Barcelona. It proved not to be as serious as some of the delays that had dogged her en route to Australia. While any deviation from a carefully planned route could be unsettling to a long-distance flyer, she was airborne again after a few minutes and heading towards Oran.

From there, her plans would take her to Gao and then Douala, off the coast of Cameroon, and Mossamedes, in what was Portuguese West Africa, and thence to the Cape. Without the benefit of a radio or parachute, she knew that any mishaps could prove fatal and would mean coming down in terrain every bit as inhospitable as those she'd endured on her previous long-distance flights. She was also conscious of the twenty-four hours of sheer hell she had put her parents through when she'd gone missing, but she was the Lone Girl Flyer. She knew what her public expected of her and she would not let them down.

The one aspect of the flight that Amy was dreading was desert flying. Her only previous experience of it had resulted in her being forced down in a violent sandstorm outside Baghdad. She had been flying for the best part of a day over the featureless desert with sand dunes sometimes up to 50 feet high. She now deftly manoeuvred *Desert Cloud*, hugging the wide muddy expanse of the Niger river towards the aerodrome at Gao.

Resting the night at Gao would have been a very shrewd decision had Amy taken it voluntarily, but after a short while in the air, she felt *Desert Cloud* seemed suddenly very light and realised that the fuel tanks had not been refilled correctly. She turned back to Gao, but by the time refuelling had been completed, the moonlight had all but vanished. She had no option but to stay the night.

Amy took off from Douala overwrought with exhaustion and worried that the light was fading fast. She had arrived late in the afternoon, as the sun was setting – the overnight stay at Gao had been a contingency she hadn't allowed for, now time was against her.

After an hour on the ground, she had refuelled and turned *Desert Cloud* round. Then, just when she least needed it, the weather began to turn against her. There was a tropical storm brewing and thunder in the sky. With her visibility diminished, Amy admitted to being very frightened. She could barely see for most of the journey from Douala, as the weather steadily worsened. When she hit cloud, Amy climbed only to lose control, momentarily, of her battered Puss Moth, which bounced around in a sky peppered with thunder and illuminated by random flashes of lightening. Boredom from long hours of mundane textbook flying was an occupational hazard, but adverse weather always provided the greatest test of a pilot's true mettle. Amy had experienced her fair share. The wind blew her off course, and when she discovered a problem with her oil pressure, it forced her into yet another unscheduled stop at Benguela.

The long hours of flying over the Sahara had blown particles of sand into the engine. Amy had no choice but suffer a long delay while a new filter was fitted. She was exhausted and her spirits sagged. In a telegram to Jim, she admitted her disappointment, but her malaise didn't last for long. Once *Desert Cloud* was ready, she took off again, with just over half a day's flying to reach Cape Town and claim her record.

A large cheering crowd greeted Amy when she landed *Desert Cloud* at Cape Town in the afternoon of Friday 18th November. She had flown over 6,000 miles in four days and seven hours, on less sleep than most could expect in a single night. It was a remarkable achievement and the press back home felt compelled to acknowledge it as such. Amy pulled off her goggles and, amid the cheering of the crowd, she allowed herself a few moments of quiet reflection. Her feelings were a mixture of exhaustion and euphoria. The elements had tested her nerve and piloting skills, and Amy had shown that she had enough pluck to pull it off. As telegraph wires fizzled, spreading news of her triumph around the globe, flashbulbs exploded as photographers jostled with each other among the swelling crowd, eager to capture Amy's smiling face, which told newspaper editors everywhere all they needed to know, the 'Empire's Flying Sweetheart' had done it again.

Having smashed the existing solo record to the Cape set by her husband, Amy wanted to give herself the best possible chance of claiming the record for the homeward journey that was held by the Duchess of Bedford and her co-pilot, Captain Barnard. For this, a full moon was imperative, so she had to stay in Cape Town and wait. This was no great hardship during a Southern hemisphere summer, especially as it was so dark and gloomy at home in England during November. Amy was, however, desperately homesick and she confessed to missing Jim. In the four months since their nuptials, they had spent more time apart due to their individual flying commitments. Frustrating though this was, Amy knew that if they were to continue enjoying holidays on the French Riviera and living the Mayfair lifestyle – of which Bill Courtenay was so critical, but had attracted Jim like a moth to a flame – they had to continue breaking records to finance it. There were still some lucrative deals to be made, but the novelty value in long-distance flying was definitely diminishing.

Amy waited twenty-three days kicking her heels for a full moon. She climbed into the cockpit of *Desert Cloud* on 11th December waving goodbye to the crowd. She looked tanned and very relaxed, and felt quietly confident that she could set another record. More than that, though, she longed to be home again with Jim. The weather had a few questions to ask her as usual, so it took three extra days to make the homeward journey. However, at seven days and seven hours, it was still a record. An exuberant crowd had braved the chill December air in Croydon to chant "Amy! Amy! Amy!" with the same enthusiasm and genuine affection that they greeted her return from Australia. Only now, it was her publicity-eager husband Jim Mollison that the press sought an opinion from, rather than her father Will, who, along with 'Ciss', was there to see Amy's homecoming.

In a break with Johnson family tradition, however, Amy would not be joining them for Christmas in Bridlington. She and Jim had decided to spend Christmas 1932 basking in the glory of her two new records, among the celebrity set in the stylish Swiss resort of St Moritz.

Although he had been full of praise for Amy in achieving her double record to the Cape, Jim's Scottish male pride had been sufficiently affected for him to swiftly turn his attentions towards seeking a fresh challenge. Having already conquered the North Atlantic, he now wanted to fly from England to Brazil and claim a new record for the fastest crossing of the South Atlantic.

While Jim was away, Amy had public engagements to attend. With Bill Courtenay for support, she went to Crufts Dog Show. The press and avid autograph hunters turned her guided tour into a media circus, but she was still photographed with many of the dogs on show. Amy looked relaxed and at ease, taking the attention in her stride. Her recent success had pushed her popularity back to its post-Australia heights, but Amy still had half a mind on an Atlantic record of her own. In purely monetary terms, its potential value was small. Although she could shave time off the existing record set by her friend Amelia Earhart, she would still only be the second woman to fly solo across the Atlantic. Jim felt they should try something truly spectacular. A joint flight would give them a definite edge over their rivals.

Amy set off to Madeira for a pre-arranged reunion with Jim, sailing home from South America. A short holiday on the Portuguese island would give them the opportunity to enjoy some winter sunshine and to formulate their plans for an audacious bid at the long-distance record, incorporating an Atlantic crossing to New York. Thus, it would make them the first married couple ever to do the 'Water Jump'.

AMY READING MAPS

14

DISASTER AT BRIDGEPORT

Seafarer was a De Havilland 84 Dragon painted black, with its name written large in white. Jim Mollison remarked to Bill Courtenay that it looked like a flying coffin, adding that it only needed some brass handles fitted to look the real thing. Amy and Jim had placed their order for the latest model in the De Havilland range in January 1933 before their plans for a round-the-world record had been cemented. Their aircraft would be specially modified with the extra fuel tank they would need for the Atlantic crossing, and then the next leg from New York to Baghdad. Their preparations were as meticulous as ever. They had decided on a take-off point in England rather than Ireland, as Jim had favoured on his solo effort the year before.

Amy had written to her father to tell him that the flight was very important to them and, without wanting to alarm them unduly, added that it was probably her riskiest venture yet. The fuel they needed for the Atlantic crossing meant they were taking off with 450 gallons in the two tanks of their twin-engine plane. As usual, they attracted a large crowd of well-wishers and sensation-seekers at Croydon Airport on June 8th, having postponed their planned take-off by three days due to unfavourable weather. Amy was smiling but apprehensive as she and Jim gazed at the cameramen from Movietone News. The awesome reality of an Atlantic crossing was finally dawning on her, while Jim was his usual nonchalant self, managing to hide any nerves behind a demeanour of slightly amused boredom.

Jim was to take the first stint of flying while Amy sat behind. *Seafarer* hurtled down the runway gaining speed, but, suddenly, they hit an obstacle. Jim swore as *Seafarer* slewed round on its belly, the undercarriage badly damaged, and the crowd gasped in horror at what they witnessed. A fire engine rushed out to meet them. The Dragon was a potential firebomb that could engulf them in flames in a matter of minutes. For Amy, already more worried about a long-distance flight than she'd ever felt previously, the accident was the last thing her flagging confidence needed. Luckily, they were rescued from the cramped interior of *Seafarer* before the fuel could catch light. *Seafarer* sat there, a mangled symbol of their hopes dashed and a serious dent in their plans, which they could ill afford. They had chosen early June to take advantage of a full moon and now the repairs to *Seafarer* would cause them a delay that would turn into weeks.

Despite its reputation as the airport of the Empire, Jim had never been particularly impressed with Croydon Airport. The incident with the damaged undercarriage confirmed his poor opinion of it and convinced him they should find a different take-off point for their second attempt. It took three weeks for the damage on their brand-new Dragon to be repaired, and during the enforced delay they searched round for the ideal venue for take-off. Pendine Beach in Carmarthan Bay suited them both. A long stretch of firm clear sand had served Jim well at Portmarnock Strand the previous August. So, they arrived in South Wales in early July. Amy had just celebrated her thirtieth birthday, but remained pensive about another delay due to adverse weather.

She and Jim were easily the most famous people that this small seaside town had ever seen, and this brought hordes of curious autograph hunters in their wake. They became a big tourist attraction and local retailers welcomed the increase in trade, but Jim, was not used to long spells of inactivity and he grew tired of the sideshow their prolonged stay had become. They returned to London to wait at a safe distance until the weather was right.

On Friday 21st July, they finally got the weather report they'd been waiting for and flew back to Pendine Beach, with plans to take-off around midday on Saturday 22nd July. The tourists who had descended on the Welsh town to see Amy and her dashing husband knew that the flyers would be returning once the weather over the Atlantic had turned in their favour and so they had stayed, while others had come and gone. The beach boasted 7 miles of smooth even sand for Jim to gain speed and lift *Seafarer* off the ground. The crowd cheered and waved as *Seafarer* became an ever-decreasing speck in the distance, heading out towards the Atlantic.

Amy relieved Jim at the controls at regular intervals throughout the flight. There was space behind for them to sleep and to keep a regular check on fuel levels. Jim had made the brave decision to carry less fuel than the maximum 450 gallons they had on their aborted take-off at Croydon. With favourable winds over the Atlantic, he felt relatively comfortable with his decision. Having shared the flying throughout the night, they saw the sun rise. They'd been in the air for twenty-one hours. For Jim, having a co-pilot on a long-distance flight was a new experience, while Amy had made all but one of her flights solo. It meant decisions could be taken jointly; however, when opinions on strategy differed, tensions could rise. By mid-morning, Jim told Amy he had seen landfall, and briefly they turned and gazed at each other as the enormity of it sunk in. They were across; the first husband-wife team to fly the Atlantic, while Amy was the first woman to have flown it in the westerly direction. She was not, however, inclined to become too overwhelmed. Their position was confirmed when they flew over the Straits of Belle Isle, Newfoundland – only 1,200 miles from New York.

The decision to take some 30 gallons less fuel always had the potential to be a prophetic one, and so it proved – largely due to stubbornness on Jim's part, as he insisted they press on non-stop to New York. Amy wanted to play safe by coming down to refuel at Boston as the light diminished. A heated argument ensued between them, but Jim wouldn't budge. Amy submitted in frustration and Jim's ill-fated opinion that they should carry onto New York was pursued. It had dire consequences for them both. For Amy, it was a decisive moment in their relationship. Just a week short of their first anniversary, a seismic rift in their marriage had been exposed.

Jim was at the controls when the engines on *Seafarer* began to cough and splutter. Their fuel level was dangerously low and in the gathering darkness of a July evening, they had to find a safe place to land. Amy sat beside him and said nothing, not trusting herself to keep her temper. If ever there was a time when she would have longed to be proved wrong, this was surely it – but she wasn't and she knew it. The most unforgivable aspect of it all was that it had been avoidable.

Seafarer came crashing down with a thud. Amy realised they had misjudged the wind direction and were landing into it, rather than against it. Jim was momentarily blinded on impact by the sudden glare of lights being switched on. He was unconscious and had been thrown through the windscreen by the impact.

When the rescue team found them, Amy beseeched them to hurry in fear that the plane might explode, but luckily Jim had cut the ignition switch as *Seafarer* had bumped along the ground. Their new aeroplane, for which they had paid almost £3000 pounds, had split in two. They had lost their momentum, as well as a realistic chance of achieving a long-distance record. Although deeply concerned for Jim, Amy couldn't avoid blaming him for his recklessness. For the sake of his ego, and the clamour for headlines, he had put their safety at risk and they had so nearly paid the price.

From her hospital bed in Bridgeport, Connecticut, Amy – still badly shaken and in shock – was anxious that her parents be reassured they were okay. She also, somewhat typically, expressed the hope that the public back home wasn't too disappointed. As soon as they were deemed fit to be discharged, they planned to head for New York, where they had a suite booked at the fashionable Plaza Hotel.

President Franklyn D. Roosevelt had been in office for just over four months, with his social economic programme known as the 'New Deal' that aimed to lift America and its people out of the melancholic state of mind it had been in since the Wall Street Crash and the Great Depression that followed. 1932 had been the lowest point in its recent history; there were millions out of work, displaced and homeless. Throughout the campaign, and right up to the November election, his oratory had lifted the spirit of the nation and won him a landslide victory over the incumbent Republican Herbert Hoover.

Amy and Jim saw for themselves the indomitable spirit of a man who had overcome polio, as well as the sheer strength of character that had compelled him to claim the Presidency. On the last weekend in July, they were invited to join him for Sunday lunch at his family home in Hyde Park, New York State. An invitation to Sunday lunch was an honour in itself as the President liked to keep Sundays for his family. The invitation included Amelia Earhart and her husband, George Putnam, who were hosting the Mollisons at their home in Rye. The hosts consisted of the President, First Lady Eleanor Roosevelt and his mother, Sara Delano Roosevelt, who ruled the roost at Hyde Park.

A photograph was taken on that Sunday afternoon, which preserved the moment for history. Jim Mollison, his head still bandaged, stood between Amy and Amelia. The First Lady, tall and angular stood on Amelia's right-hand side, while the President, clutching his stick, stood on Amy's left. For those astute enough to notice, there was still some residual tension between Amy and Jim.

Two days later, on August 1st, they were given a ticker-tape parade through Manhattan. Amy and Jim sat high in the back of an open-top car as the procession meandered its way through the concrete canyons of Manhattan to a civic reception at City Hall. Amy waved and the crowd cheered enthusiastically, as ticker-tape rained down on them. It had become a tradition for New York to put on something spectacular for aviation heroes and Amy could bask in the afterglow of victory, just as Jim had done after his solo effort the year before. Having 200,000 jubilant people come out to greet them on a scorching summer day was confirmation of their achievement in crossing the Atlantic together, despite the less than auspicious nature of their arrival.

Amy decided to remain in America, while Jim sailed home on the *Majestic* to acquire a new aeroplane to replace *Seafarer*, which, in addition to the damage it had suffered in the crash, had since been stripped bare by trophy hunters, who descended on the crash site at Bridgeport like a horde of vultures feeding off a rotting carcass. Amy retained the suite at the Plaza and spent many hours with Amelia. She felt much more relaxed in Jim's absence and visited the Boston-Maine Airway, a small airline in which Amelia Earhart held a financial stake. Her friend was much more independent in her marriage to George Putnam, whose business interests in publishing didn't impinge on his wife's. Thus, there wasn't any of the inter-dependency, the competitive rivalry or jealousy that was becoming ever more apparent in Amy's relationship with Jim. Amy also grabbed the opportunity to go flying. She flew for Boston-Maine Airways and briefly was a co-pilot for TWA – the airline that later attracted the attention

"TICKER-TAPE" ON BROADWAY

of the obsessive multi-millionaire aviation innovator Howard Hughes, who, by the end of the 1930s, had acquired 25 per cent of TWA and would eventually control almost 80 per cent of the company.

When Jim returned, by ship, with a new aeroplane that they named *Seafarer II*, Amy met him at Union Station, Toronto. They had plans to resume their bid for a long-distance record and had chosen Wassaga Beach, Ontario, as their take-off point. However, after two aborted attempts, they decided to abandon the whole project. Their bid to set a round-the-world-record was a valid and noble enterprise, but their experience at Bridgeport had ultimately sapped their enthusiasm and now the momentum was lost.

Their shared near-death experience had certainly changed them, but not in a positive way. A serious fault had appeared in their relationship after that fateful night in July. Although there were other aspects of Jim's behaviour that Amy found hard to take, and which would ultimately help to sour the romance of the 'Flying Sweethearts', their marriage changed after Bridgeport and it never recovered its old magic.

To add to her woes, Amy suffered a serious fall when riding and a routine check-up revealed another underlying health problem that required a prolonged rest. Amy was frustrated but felt resigned to her fate, while Jim dissolved into one of his sulks and went to Bermuda alone. Amy promised to join him as soon as she felt strong enough to travel. In the meantime, she rested in her suite at the Plaza Hotel at the largesse of its management.

Rumours of a rift between the couple had been circulating in the British press for some time, but Jim strongly refuted any suggestion that all was not well. He had described the holiday in Bermuda as a 'second honeymoon' but, to the casual observer, they were far from a honeymoon frame of mind. There was a distinct coolness between them when Amy finally joined him.

Jim was drinking heavily, as usual, and his overall behaviour was as intolerable as it had ever been. By mid-November, Jim returned to England alone, while Amy went to Florida, remaining there for the rest of the year.

15

THEIR LAST HURRAH!

The MacRobertson Air Race was the kind of event that could still attract the public's attention. Although headlines about the latest solo record were growing tedious, as aviation became more commonplace with scheduled services being established all over the world, a race to be the fastest pilot from England to Australia with a cash prize of £10,000 was enough to bring together the most accomplished pilots of the golden age of aviation. Jim had cabled Amy in America to ask if she was interested in entering and to committing herself financially to the joint purchase of a new aeroplane. Amy took some convincing. It was a dilemma for her. She didn't particularly relish the thought of another flight with Jim, but if she had sought another partner, the press who enjoyed speculating about the state of the Mollison marriage would have had a field day.

Amy was busy evaluating her life and weighing up various options. She was becoming bored with the nomadic existence albeit in a lavish hotel suite that she led while Jim's fascination with the Mayfair Set showed no sign of waning. Amy yearned for something different. Her place in history and in the public conscience was shifting by the time 1933 faded out, and the job in commercial aviation that she'd once wanted now seemed possible. She craved privacy over notoriety, and employing Amy Johnson didn't seem as far-fetched as it would have done after her return from Australia.

Having decided to accede to Jim's desire to enter the MacRobertson Air Race as a husband-wife team, she set about establishing her own plans for well-paid work. Through Bill Courtenay, she secured a six-month contract as Aviation Editor on the *Daily Mail* and, while in America, visited the Beechcraft factory in Kansas, hoping to secure a licence to sell their aircraft back home. Amy craved mental stimulation and wanted to explore every possibility of improving her earning potential, leaving Jim to his own devices and the life of idleness he seemed to thrive on.

Sir Macpherson Robertson, a wealthy Australian entrepreneur and chocolate maker, who sponsored the Air Race giving it his name and moral support, was determined that his beloved Melbourne would be the finishing line. As the race was being held to strengthen ties between Australia and the Mother Country, it had been decided to start the race in England – and so began the process of finding a suitable venue.

The most obvious choices were deliberately overlooked, and to great surprise and considerable derision, a sleepy backwater of Suffolk that had just opened its new air base was selected. So, RAF Mildenhall made its first bow to the world.

The De Havilland Comet, which cost Amy and Jim £5,000, was one of three in the race. They called it *Black Magic* – a somewhat curious choice, given the nickname Jim had foisted upon their Dragon for its black livery. Another of the *Comets* was being sponsored by the management of the Grosvenor House Hotel and was carrying its name, with Charles Scott and Tom Campbell-Black as the pilots – a decision that must have left Amy and Jim piqued as this was the hotel they had called home.

There were two aspects to the race and a cash prize would be awarded to the winner of each. The winner of the speed race would receive £10,000, while the winner of the handicap race would net £2000. Race rules dictated that no team could keep both cash prizes, and the handicap race would be decided upon a set criteria such as engine power and aircraft weight. The surest way of winning a cash prize was to be the first team to land at Melbourne, and so a ballot was held to determine which team took off first. Fortuitously, the Mollisons won the ballot. Amy was acutely aware that they had been given a considerable advantage with the draw and they could not afford to throw it away.

Another fascinating aspect of the Air Race, which organisers were keen to exploit for its publicity value was the level of fierce competitive rivalry that existed between the teams. Amy remembered Charles Scott well from Australia, and his views on women pilots was very well documented. He basked in his chauvinistic reputation unashamedly, while he and Jim had also crossed swords before. Over time, they had developed a mutual dislike of each other that was close to loathing. Scott's flying partner was Tom Campbell-Black, another dashing pilot whose womanising ways were as notorious as those of Scott and Mollison. He was courting the actress Florence Desmond at the time of the race, but his other lovers had included Beryl Markham, who, in addition to being the first woman to hold a racehorse trainer's licence in Kenya, had flown the Atlantic chronicling her adventures in the bestseller *West With the Night* – although there was some doubt over her authorship of the book and the controversial rumours dogged her for many years. Beryl Markham could boast that one of her many lovers was Harry Duke of Gloucester, and Amy would often bump into him on the back stairs of the Grosvenor House when he was arriving for an assignation with Beryl and she was escaping the autograph-hunters who consistently waited for her at the hotel's entrance.

Not all the competitors approved of Mildenhall as the starting venue. At first, their curiosity was aroused by the choice of an unproven venue for such a prestigious event. Adequate accommodation of a standard to suit the ego of a record-breaking flyer was minimal in the area, and basic facilities at the base were deemed 'spartan' thus many teams delayed their arrival in Suffolk for as long as they could.

At dawn on the third Saturday in October 1934, the first team Amy and Jim in their Comet *Black Magic* were called to the starting position, at approximately 6.30am, by an official race marshal. The most lucrative "Grand Prix of the Skies" ever to be held was about to start. After *Black Magic*, the teams would take-off at forty-five second intervals and head into the autumn sky towards the first stopover in the Iraqi capital, where an official race marshal would record their arrival.

Each competitor felt the rush of adrenalin as they prepared to push themselves and their aircraft to the absolute limits. All shared one common goal: to be the first to land at Melbourne and claim the spoils of victory. The race was dramatic from the start, as the teams pushed themselves on for a stab at glory. Perhaps inevitably when the stakes are so high and the prize so great, there was to be tragedy.

For two New Zealand pilots, the race and their lives ended when their plane crashed on landing in Southern Italy, bursting into flames before the airmen could be rescued. Charles Scott and Campbell-Black had drawn sixth place in the ballot and they ran into bad weather. More drama unfolded as the race continued towards Baghdad.

Amy and Jim remained on the ground at Baghdad for just as long as they needed to. They were relieved that *Black Magic* had been the first plane to reach the first of five mandatory stops. They learnt later that they had been the only team to get there non-stop from Mildenhall. Amy went for some light refreshment and a hot bath, while Jim inevitably headed for the bar. They were both ecstatic to be taking off again before any of their competitors, Scott and Campbell-Black, in particular had even touched down in Baghdad.

The second mandatory stopover was at Allahabad, and Amy knew that if they could reach it non-stop their chances of being the first team to land in Melbourne would be pretty good. In spite of having to stop at Karachi, their confidence was still high. Jackie Cochran was forced to retire when her plane was damaged after landing, and it meant that Amy now held the distinction of being the last female pilot in the Air Race. For this distinction alone, Cochran wished her the best of luck. Amy was always as fiercely competitive as the men and was keen to put one over Charles Scott after he'd been so cruel to her in Australia. In light of the tragedies already suffered, though, she urged caution on Jim and could only hope that this time he would heed her advice.

Had Amy started to believe that their flight was going too well, her instincts would have been sound – for it all began to unravel after their first take-off from Karachi. Jim's poor judgement forced him to return to Karachi, believing wrongly that there was a problem with the *Comet's* retractable under-carriage. Amy took the controls when they were airborne again and hit bad weather. She forced herself to wake Jim so he could assist her with the navigation. Her fragile patience was stretched further when he realised they'd brought the wrong maps. Rather than risk a night landing at Allahabad, they opted to return to Karachi again, thus losing more precious time.

When *Black Magic* left Karachi for Allahabad, fortune turned against them again. There was thunder in the air, and the electricity it generated played havoc with their compasses. Soon, tempers began to fray, and when they made another unscheduled stop at Jabalpur, they discovered there was insufficient aviation fuel available. Jim instructed that low-grade fuel be pumped into *Black Magic's* tanks to get them to Allahabad, but Amy was against the idea. She knew from experience the damage it would do to their engines, but Jim insisted they had no other choice. When they took off for Allahabad, the Comet was not handling well. Amy was in the pilot's seat when smoke suddenly appeared from one of the engines and she knew they were in deep trouble. When the engine failed, Amy expected they would have to abandon the Comet, but she somehow made it to Allahabad and was furious that her advice had been ignored yet again. All the doubts that she'd had about flying with Jim quickly resurfaced.

Amy was in tears by the time she showed her face. She had nothing left to say to Jim and the animosity between them was tangible. *Black Magic* was going nowhere. It needed a replacement cylinder, which would have to come from Baghdad. The knowledge that Charles Scott and Tom Campbell-Black were on their way to Singapore, then onwards to Darwin and Melbourne – and the finishing line, which they eventually reached early on Tuesday morning only served to plunge the Mollison's spirits even further. Between the four of them, the rivalry had been intense. – A race within the race, in which pride was at stake. Their bright red Comet, named *Grosvenor House*, had won. They also won the handicap race and had reached Australia from England in just over four days, cutting out Europe altogether. There was scant consolation for Amy and Jim, who were still furious and barely speaking. They watched the stragglers land at Allahabad, refuel and fly off again, leaving them stranded. They had nothing to do but wait and confront the reality that they had failed, and it hurt.

It took three weeks to repair *Black Magic* before they could fly it home. They took a short break in Calcutta, but there was no sign of a thaw in their relationship. Amy felt no compulsion to forgive Jim this time, unless he demonstrated a willingness to change his ways. The future did not bode well for them.

Amy dreaded the long winter in their lavish suite at the Grosvenor, if this was the pattern of behaviour that Jim intended to follow.

She wanted to spend Christmas in Bridlington, but Jim preferred Switzerland. So, to avoid any more unpleasant stories appearing in the press, and as a conciliatory gesture, Amy agreed. It was yet another decision that she quickly came to regret. Amy's life had reached a definite crossroads as 1934 came to a close. Her marriage to Jim was the crux of her problems, but she also had concerns professionally. In some respects, the MacRobertson Air Race had been an epoch-defining event. Its success had increased the clamour for the establishment of a scheduled air service between England and Australia. It had also been a definite line in the sand. The coverage it had received in the press and the praise heaped upon the victors served as a swansong to a diminishing age.

Pilots like herself, Jim, Amelia Earhart and Charles Lindbergh, among many others, had proved that records could be set and broken at will between any two compass points on the globe. Their work was almost done. The future looked uncertain for the record-breaking flyer at the start of 1935. However, Amy was sure of one fact: the age of the "Flying Sweethearts" had seen its last hurrah.

The atmosphere between them soured the more Jim drank, and on return from Switzerland, he abruptly left Amy again for America. Amy, at Will's behest, went on a riding holiday to the Malvern hills with her youngest sibling, Betty, where a deep understanding developed between them and confidences were shared. Amy felt able to unburden herself and Betty, now almost sixteen, felt she'd found a kindred spirit in Amy, who was much older and had once seemed so distant. Although the experience hadn't been entirely cathartic – and it had probably done Betty more good than her – Amy returned to London and took up residence at the Savoy. Jim was resolved to remain where he was and found the single life suited him. However, Amy was sensitive to gossip in the press, so she pleaded with him to move in with her, which he agreed to do.

Amy took a trip to Madeira, but her listlessness and a distinct lack of purpose in life frustrated her. She had few good friends, but an abundance of casual acquaintances, and she longed for a resolution to the endless stalemate her marriage had become. One female acquaintance with whom Amy had formed a slight attachment was Diana Caldwell – an elegant blonde, who, like her namesake Diana Mitford, was considered one of the most beautiful women of her generation. She ran a fashionable cocktail bar called the *Blue Goose* just off Bond Street in the heart of Mayfair, and she had a passion for flying. Amy lent Diana her plane for a jaunt to Vienna, where she had one of her numerous lovers. Her love life became a byword for scandal after her marriage to the much older and wealthy Jock Broughton. Through him, she met the man she would later describe as the great love of her life, Lord Joss Erroll, who was famously murdered in Kenya's notorious Happy Valley shooting of 1941.

Jim provided Amy with the motive she needed to end their marriage, albeit inadvertently and in the cruellest way imaginable. Amy returned to their suite at the Savoy earlier than expected one night to find Jim with a woman, both hopelessly drunk and asleep on her bed. While she had always been suspicions of Jim's infidelity, she was now confronted with the stark reality of it. And the fact that he showed so little contrition only served to compound her misery.

Amy left the hotel to stay with friends, and although she made no immediate move towards the formality of divorce proceedings, she believed there was nothing left between them that was worth saving. As the Mollison marriage continued to unravel, Jim pursued his various conquests openly – no longer burdened by any compulsion to spare Amy's feelings. Prominent among these were the actress Dorothy Ward and Beryl Markham. Amy endeavoured to retain her dignity, even though Jim's infidelity had been the worst kept secret among the Mayfair Set and had been going on almost from the start of their marriage.

FLYING - MR. & MRS. J. A. MOLLISON

AMY AND JIM

Photo © East Riding of Yorkshire Council

16

THE WILDERNESS YEARS

As Jim's influence over Amy diminished and they led increasingly separate lives, Amy's old friends emerged once more to offer her their moral support. If she ever needed an emotional crutch to lean on, Peter Reiss, who Amy had known for some years, was more than happy to provide it. Many of her friends had discreetly retreated from the scene after receiving a verbal lashing from Jim. He could be particularly spiteful when drunk and would routinely hurl abuse at Amy, and anyone else who happened to be there at the time.

Jim visited Amy occasionally at her rented cottage on Upper Belgrave Street and attempted to use his considerable charm to dissuade her from seeking the divorce that she desperately wanted – and which she believed was inevitable. Jim knew that he would fare worse from the fall-out of a divorce and yet his behaviour didn't change. He still drank excessively and he chased women with the carefree attitude of a single man.

Bill Courtenay was still very close to Amy and he was willing to offer his advice about various business ideas she was mulling over, but his personal circumstances had changed dramatically. He had suffered a personal bereavement, and private matters demanded more of his time. These factors combined had forced him to drive a much harder bargain with Amy for the cost of his expertise. Amy was unsure of what she could reasonably earn and so she was unable to pay Courtenay the higher fees he was seeking. There had been suggestions that she turn her hand to writing seriously and any publisher in 1935 would have paid handsomely for the Amy Johnson autobiography or even for her first novel, but writing had never come easily to her, even her regular articles as Aviation Editor for the *Daily Mail* had been a hard slog. The prospect of being hunched over a typewriter, suffering from writer's block, did not appeal to Amy at all. She longed for the freedom to fly.

Air Cruises combined Amy's need to earn money with her passion for flying. She would fly tourists to the most chic and cosmopolitan cities of Europe. Paris, in particular, appealed to her – a city for which she had developed a genuine love – but also Rome, Vienna, Madrid and Berlin, which was to host the Summer Olympics in 1936. She had gained the interest of a firm of luxury car dealers who operated out of Bruton Street, just off Berkley Square, and their prestigious address alone would add a certain cachet to a fledging business like Air Cruises. She also had the moral support of her father, with whom she was corresponding regularly again. Will's business acumen had always been sound and his enthusiasm for the idea lifted Amy's spirits. In a letter she wrote to him early in 1936, she told him about another potential backer, who she referred to simply as F. D. a wealthy middle-aged Frenchman, with whom she had become friends. He had developed a passion for flying and owned his own aeroplane. The jewel in his expansive business empire, however, had to be the opulent Georges V Hotel in Paris – one of many he owned in Europe and North America. He was an entrepreneur in the true French meaning. Amy had every reason to believe that fortune was smiling upon her once more after a wretched period of bad luck, for she could not have found a better champion than Francois Dupré.

The Air Cruises venture was launched with considerable fanfare in March 1936. Amy had appointed Bill Courtenay as her publicity manager. She was also seriously considering something that she had believed was off her agenda since Mildenhall: a stab at improving her 1932 solo record to Cape Town and back, with the specific purpose of publicising her new business. Amy wanted to keep her plans secret from the press as much as possible. She had needed them to publicise Air Cruises, but they'd still been more intrigued by the conspicuous absence of Jim. Amy was so tired with details of her private life being treated as front-page fodder.

Her flight plans were as meticulous as ever, and she was setting off in a Percival Gull cabin monoplane on 3rd April 1936. For the first time, she'd be carrying a radio. In another break with tradition, she'd chosen to take-off from Gravesend Airport, and to mark the flight – her first solo effort for almost four years – Amy's favourite couturier, Elsa Schiaparelli, had designed a navy blue suit and silk scarf with a curious newspaper print pattern. Amy wore it for pre-flight publicity shots in a strangely stilted studio pose. She was still very much sought after by couturiers like Schiaparelli and Chanel, even though the public's fascination with pioneer flyers was supposed to be on the wane. Indeed, many women still wanted to follow the fashions favoured by Amy Johnson. There were other notable women in society setting the trends and none more so than an elegant American divorcée named Wallis Simpson. Her infatuation with Edward, Prince of Wales, which hitherto had been kept out of the Press would lead to the Abdication Crisis later that year, when he sacrificed his throne and the Empire for the woman he loved.

Will, her sister, Mollie, and brother-in-law, Trevor Jones, were there to see her off. There was also Jack Humphreys, who had heeded her request and discreetly disappeared from her life after her engagement to Jim. This time, however, there was no Jim by her side – although at Courtenay's behest, she had cabled him in Australia to inform him of her plans, just in case a mischievous reporter sought him out for a quote. When he cabled back to say he'd try to get there to see her off, Amy thought ruefully how typical that was of him. Being an attentive husband was not one of Jim Mollison's natural characteristics, so Amy knew that his desire to be there was proof that publicity for himself was foremost in his mind.

With a casual wave goodbye, a confident and smiling Amy climbed into the cockpit of her Percival Gull and headed off. Making good progress, she reached Colomb Bechar in the blistering heat of mid-afternoon, but then achieved a quick turnaround. She had done this before and was optimistic about her chances of setting a new Cape record. She climbed back into the cockpit for a night-time flight across the desert, but the runway at Colomb Bechar was in a poor state. Gaining speed, Amy lost control as her Gull, hindered by winds, veered to the left. She heard a tearing crash and knew instinctively that she was going nowhere.

The Gull was damaged and, with a full tank of fuel, Amy quickly abandoned the plane before it exploded. She was angry and very frustrated, but there were no tears shed this time. The crash had been due to factors that were beyond her control. She was known for her pluck and would go again.

After a wait of almost a week, followed by other logistical problems in getting the Gull repaired, Amy was back at Gravesend on 4th May 1936 – almost six years to the day she had stood on the tarmac at Croydon beside her Gypsy Moth, *Jason*, waiting to fly off to Australia as a virtual unknown. How she would have loved that anonymity now, Amy thought ruefully as the photographers' flashbulbs exploded and the Fleet Street hacks fired a volley of questions at her, agog at the sight of Jim standing beside her. Their sense of surprise only exacerbated when Jim took Amy in his arms for a passionate embrace. It suited Amy to have Jim with her, even if the ever-loyal Peter Reiss was upset by his presence. It kept the press focused on her flight, while reminding them pointedly that their wildly speculative reports that the Mollison marriage was doomed had proved woefully premature.

Amy experienced no problems on her second attempt at a new Cape record. Landing on 7th May three days and six hours after taking off from Gravesend she smashed a record that had been set by Tommy Rose as recently as February. She wasted no time on her return flight and came back via Johannesburg and Cairo, in triumph, in just over four days. Amy was ecstatic. The 1936 double record surpassed her earlier effort in terms of its significance. Amy's flying career had been in the doldrums. She hadn't completed a successful solo flight since November 1932, and the negative headlines that she had received as one half of the "Flying Sweethearts" after the disasters at Bridgeport and Allahabad had allowed other pilots, notably Jean Batten, to overtake her. However, with a bulging bag of fan mail, a deal for the exclusive rights to her story sealed with the *Daily Express* and a double record – with the outbound leg smashed by eleven hours – the front page headlines were hers once more. The Lone Girl Flyer was back.

In addition to the money that she earned from the deal with the *Daily Express*, Amy also received a cheque from Lord Wakefield for her continued endorsement of Castrol Oils. Her publicity boosted the profile of Air Cruises. Amy had every reason to be content. She had consented to Jim's pleas for another chance to make their marriage work, but it would be on her terms, and on the provision that he remain faithful. He would have to prove himself this time, and Jim's smiling face and easygoing demeanour when he accompanied Amy on an engagement to open Butlins at Skegness showed he was making an effort. Jim had gone to see Amy at her rented property in Ennismore Gardens, Kensington; where the household consisted of Amy, her housekeeper, her secretary and her pet dog, an excitable Dachshund called Teena.

The reconciliation with Jim, however, was sadly short-lived. On holiday in Juan-le-Pins, his behaviour was as intolerable as before. Amy had endured enough. All their grandiose plans to jointly cash in on the success of her recent flights amounted to nothing, as Jim demonstrated to Amy that he simply couldn't change his ways. With her confidence renewed, she didn't sulk or resort to tears. She simply told him that she was finished trying. When asked directly by reporters outside her home later that year, she told them that she and Jim had separated. There would be no more pretence for appearance's sake. It was over.

François Dupré had never made any secret of his affection for Amy. Even while Jim was in Australia, and before their ill-fated reconciliation, the French businessman was attentive, generous and totally focussed in his courtship of Britain's most adored aviatrix. He had put a suite at the George V Hotel at her disposal and demonstrated a Gallic tendency for grand romantic gestures. He saw his considerable wealth as a trump card in his pursuit of a beautiful woman. His life would always be dominated by work, but his leisure interests were varied. Aside from flying, he had a keen interest in breeding racehorses, and his stud-farm at Pont-d'Ouilly in the rolling French countryside was impressive. He was very dapper and charming, and had developed a taste for fine tailoring, but he was also somewhat fastidious in his demand for neatness and order. Nonetheless, François could give any woman a life of indulgent luxury. Since divorcing his American wife, a member of the dynastic Singer sewing machine empire, he was seeking a new consort. Amy had never showed any romantic inclination for Dupré. The relationship for her was based on friendship and a shared interest in establishing Air Cruises. However, he told her in a cable sent from Canada that he missed her and that his feelings for her were genuine. François Dupré would do whatever it took to have Amy by his side.

Beryl Markham also intrigued François Dupré, as she had intrigued many other men with her bewitching charm and beauty. While she couldn't claim the place in his affections that Amy had, he was sufficiently intrigued by her skills as a pilot to offer her a job with Air Cruises. She was recommended to him by her estranged husband, Mansfield Markham, a friend of Dupré. Amy, meanwhile, who Dupré had urged for some time to divorce Jim and commit herself fully to him, was dragging her feet.

Beryl's association with Air Cruises was brief, however, for when Lord Carberry challenged her to fly the Atlantic in his Vega Percival Gull, she accepted gladly. In September 1936, Beryl Markham achieved the distinction of being the first woman to fly the Atlantic in the westerly direction alone from a take-off point in England. Jim Mollison flew down to Abingdon, Oxfordshire with Beryl and lent her his wristwatch as a good luck symbol. When Amy heard about this, she was furious. It was an inappropriate gesture for a married man supposedly giving his marriage a second chance. It proved to be the trigger that finally prompted her to file for divorce.

Beryl's decision to go for trans-Atlantic glory meant Air Cruises needed a pilot, so Amy returned to the business venture that she had abandoned and was back where François Dupré wanted her. The problem for Amy was that she had never been so avidly wooed before meeting François, and although it was exciting, it was also a little unsettling. She had always been confident in her attractiveness to men and when she was at university in Sheffield, her love of dancing had prompted her to seek out good-looking partners whose competence and enthusiasm on the dance floor matched her own. She had continued the practice during those early weeks in London, when the bright lights of Piccadilly had dazzled her, but she was only ever lending them her body never her heart. That had belonged to her errant Swiss knight Hans, whose reluctance to commit had always infuriated her.

Later, her engagement to Jim had come about so quickly and was so exhilarating that her feet had barely touched the ground. She had known the spark was there, but with François it was sadly missing. There was no passion for her, so to be told of his was a shock and one that she didn't want to have to confront.

François took Amy to Pont D'Ouilly for the weekend, when it was bathed in autumnal splendour, and he suggested they acquire a new aircraft for Air Cruises. He also proudly escorted her around the best couture houses in Paris, so she could add to her wardrobe. As charm-offensives went, it was truly impressive. Had he been sealing a business deal, his efforts would have proved successful, but this was Amy's life. Her future. If nothing else, she needed time. Their relationship meandered on, but François Dupré was not yet ready to abandon hope.

Amy sailed to America in March 1937, the pattern of her life having settled, and she divided her time between London and Paris. She had come to prefer life in the French capital, not only because of the opulent surroundings, but she also enjoyed her suite at the George V Hotel. François was very patient. He was prepared to play the long game in his pursuit of Amy and she was content to let matters drift. However, she was conscious that someday he might force her hand – probably by proposing marriage – and then she would be forced to make a choice. She had managed to secure an intensive course of navigation at the US Naval Academy in Annapolis, Maryland, but the main purpose of her visit was personal. She had appointments with her dentist, and with Gloria Bristol, a friend who was seeing Amy in her professional capacity as a beautician, as well as playing hostess at her apartment on Fifth Avenue.

Amy was fretting over her finances again. Her bank balance was dwindling worryingly and although she had shared her concerns with Will, she was determined to demonstrate outwardly that all was well. She arrived at the gangplank in a smart new outfit that had been designed for her by Mainbocher. Amy was feeling a little out of sorts as she boarded the liner at Cherbourg. The crash she had suffered the previous October, when she had tried to land in fog at Chelsfield Kent, had shaken her badly, and Jim seemed intent on making life difficult for her in regards to their divorce. Still, she had severed one aspect of that period of her life by deed poll. She had changed her surname back to Johnson.

Whether or not François Dupré used the time that Amy was away to reassess the situation in regard to his courtship of her, or simply to evaluate the strength of his feelings is unknown, but Amy sensed a distinct change in his mood towards her once she returned to Paris. There had been tension earlier in the year when he'd made the difficult decision to wind up Air Cruises, but the coolness could not be ignored. She decided to move out of the George V and acquired an apartment instead – hotel living had begun to grate, just as it had done in London. François Dupré would still invite her to Pont D'Ouilly occasionally and insist that she stay the weekend, but another dispute erupted between them over an aeroplane that she refused to share the cost of and the old tensions resurfaced. Amy knew there was no future for them, and that the time had come for her to sever ties and walk away. Will urged her in a letter not to make a hasty decision, reminding Amy that François Dupré was offering her a lifestyle that she had become accustomed to and would now be unable to sustain by her own means. Amy knew what a gilded life she was walking away from. She would never have to fret over her finances again, but she had enough self-respect to know she couldn't settle for being a trophy mistress and that she would never be happy as his wife.

François Dupré finally found a wife in a dark-haired Hungarian named Ana Nagy, who was not yet thirty when their engagement was announced. She remained with Dupré for the rest of his life and her loyalty over nearly thirty years of marriage netted her millions.

Amy was deeply shocked and upset by the death of Amelia Earhart. The friendship that she had forged with the tall, tousle-haired American aviatrix-cum-social campaigner was the most enduring she'd ever enjoyed with a fellow flyer. Amelia and her navigator, Fred Noonan, were lost over the Pacific in July 1937, during their bid to circumnavigate the globe, an idea that had once interested Amy. She felt the same sense of numbness at the loss of Amelia as she had when the Airship R101 had crashed seven years before, claiming the life of Sir Sefton Brancker among others. Amy dealt with the news by retreating further from the public gaze. Life in London appealed to her less than ever. Her life was in hiatus and she needed time to think. She rented Dragontail Cottage in Haddenham and later took a timber-beamed cottage, named Monks Staithe, at Princes Risborough. Jessamine, her housekeeper from Ennismore Gardens, joined her there, as did her beloved Dachshund, Teena.

The publication that year of *Playboy of the Air* by her estranged husband, Jim Mollison, added to Amy's woes. The book sensationalised Jim's worse faults and he seemed to bask in the glory of the public's shocked reaction to it. His decision to dedicate the book to actress Dorothy Ward – after whom he had already named one of his planes – was seen as a deliberate jibe at Amy. Amy viewed the spectacle from her new home in the tranquillity of the Chiltern hills with regret that Jim's constant need for attention would lead him to sacrifice his self-respect for so little transient glory.

Amy was still desperate to boost her income, and although selling her Beechcraft aeroplane brought in some money, she was compelled to accept a loan from Will. The money worries never vanished completely, though, or for very long, and while writing brought her some much-needed funds, it always seemed to Amy like a lot of hard slog for comparatively little reward.

She did contribute to *Myself When Young*, a series of essays by prominent women of their day, and she later published *Skyroads of the World*, a book about expanding commercial aviation and the establishment of scheduled air-routes crisscrossing each other all over the world – covering many of the destinations in which she had set and broken records. However, the autobiography that many had anticipated from Amy in 1930, after she had returned from Australia, never materialised.

By the end of summer 1938, Amy was finally a free woman as her divorce was decreed absolute. A return to life in London and the Mayfair Set still had no appeal for her, and life in the country had given her a new lease of life. Concerns over money continued to trouble her, but she'd taken up gliding. Soaring into the air above the Chiltern Hills when the wind was favourable gave her the same sense of liberation she had once enjoyed in an open-cockpit biplane.

In November 1938, Jim Mollison married Phyllis Hussey, a wealthy divorcée, in a civil ceremony at Caxton Hall Registry Office. The news came as no surprise to the press, who had always liked Jim, despite his many failings, as he always provided good copy. How Amy reacted to the fact that Jim was beginning his second marital adventure a mere eighty days after theirs had been formally ended by law is unknown. All she could do is hope that the second Mrs Mollison fared better than the first.

Amy's casual acquaintances, who styled themselves as archetypal members of the Mayfair Set, had grown used to seeing her out and about town during the years she was at the height of her fame. They looked at the rural life she was living now and dismissed it scathingly as a "wilderness" existence. Amy was not hurt by their harsh analysis, because it was a "wilderness" of her own choosing.

17

THE ATA: A NATION AT WAR

In 1939, Amy was left shaken and traumatised by a car accident, in which she sustained injuries more severe than any she had ever experienced before. The worst of them was a broken right knee, which left her immobile and in great pain for weeks. The accident occurred in February, at Witney, Oxfordshire, and came at a time when she had been driving competitively in car rallies for fifteen months. In the previous month, Amy had driven in the Monte Carlo Rally, partnering Dorothy McEvoy.

The idleness forced upon Amy by her injuries left her feeling morose, and she soon sailed for America on board the famous liner *Ile de France*. She looked fragile, tired and unhappy. The sparkle had gone from her blue eyes and the passengers that did recognise her as the girl who had flown to Australia, and then became the most famous woman in England on the strength of that achievement, saw someone whose star was in descent. Amy lacked direction and she longed to have a purpose in life again. Sat on the deck of the luxury liner in her wheelchair, she was pale and drawn. She craved anonymity. On arrival, Amy gazed at Liberty's greenish tint in the light of the early spring day, as she had on many previous Atlantic crossings. She would not have believed she was seeing the famous old statue for the last time.

Amy had several appointments in New York, aside from the one with her dentist. As always, she had one eye on potential business opportunities. She had unfinished business with Gloria Bristol that she wasn't looking forward to resolving, as the Park Avenue beautician had muddied the waters between them over unpaid bills. Amy was furious that she had involved François Dupré in her private affairs, by showing him letters that had passed between the two of them. It was a salutary lesson for Amy, not to mesh friendship with business. It was not an error that she would be making again. She told Gloria Bristol in a letter what she thought of her behaviour in strongly worded terms, thus ending their friendship.

The tense political situation in Europe didn't help matters either. There was a general sense of unease; just as the nation was emerging a little stronger after the long years of austerity during the Depression, there was yet another challenge to face. The German Chancellor Adolf Hitler had annexed his homeland Austria the previous year, and this had come after Germany's claim on the Sudetenland had been realised with apparent ease. The fragile "Peace in Our Time" that Prime Minister Neville Chamberlain had secured at Munich the previous October now reassured nobody, and the belief that a conflict between Britain and Germany in the not-too-distant future was now almost inevitable became the dismal view shared by many.

Amy remembered watching the Zeppelin raids over her home in Hull in 1915 as an adolescent, wide-eyed with wonder, but with no real concept of what she was witnessing. It wasn't until they lost their uncle, Bert Johnson, followed by another relative, Hamlyn Petrie, that the harsh reality hit home. Now, as the dark clouds over Europe became ever more menacing, the possibility that the world could be plunged into yet another apocalyptic conflict – just two decades after the war to end all wars – was inconceivable to the ordinary man and woman on the street.

Amy returned to England in April, and although the trip had done her good, it hadn't entirely lifted her out of her gloom. However, within a month, she was pleased to have secured a flying job with the Portsmouth, Southsea and Isle of Wight Aviation Company. The salary of £1.00 a day, plus 10 shillings per hour for flying, was not inspiring, and paltry compared to what she'd been used to getting for selling the story of her latest flying exploit. Nonetheless, it would do her for the time being. It was the first time in nearly ten years that Amy had been engaged in regular salaried employment, when she had given up her a job as Vernon Wood's secretary to work in the hangers at Stag Lane. Amy still wanted better paid work, but the ferry job gave her a purpose, lifting her out of the malaise she had been in. It allowed her to believe her life had turned a corner at last.

Amy adjusted to the routine of her job, although the pattern of each day varied from the previous one, and any night-time flying was an opportunity to earn more money. She was corresponding regularly with Will and they shared their opinions on the tense political situation, as well as the likely outcome. By August, fears of war had heightened. On Friday 1st September, German tanks rolled into Poland and a weekend of feverish diplomatic brinkmanship ensued. Sadly, at 11.10am on Sunday 3rd September, Neville Chamberlain delivered a radio broadcast to the nation. With a voice leaden with regret, he announced that Britain was at war with Germany.

Amy was justifiably concerned that the declaration of war would have an impact on her new job, and the irony of it could not have escaped her. After years of patiently waiting for her first proper flying job, one had finally come her way, only to be affected by the crazed jingoistic ambitions of an Austrian-born tyrant named Hitler.

P. S. & IOW Aviation had secured army contracts before the war had begun. Amy had piloted a wireless operator on a training exercise and had taken a senior officer to France, but these assignments were sporadic and the majority of her ferry work was predominantly mundane flying. On the day of the German invasion, Amy had been told to report to Heston Airport, while other staff were relocated to Cardiff. After much work and negotiations behind the scenes, involving the Minister for Civil Aviation Sir Francis Shelmardine, a plan came to fruition to utilise the skills of pilots, deemed no longer suitable for regular RAF service, in the interests of the war effort. On 3rd September the day of the declaration, Gerald d'Erlanger was appointed Commanding Officer of the Air Transport Auxiliary.

Amy's orders to report to Heston amounted to nothing more than clerical error, and she was redirected to Cardiff with the rest of her P. S. & IOW colleagues. The first months of the conflict, known as the 'Phoney War', were largely uneventful, although the Luftwaffe started their bombing campaign in the first month and a mass evacuation of children from London and other major cities began in earnest.

The capital was sandbagged and strict black-out regulations were put in force, as the nation, which for months had hoped to avoid the possibility of war, now grew accustomed to its reality. Amy continued flying for P. S. & IOW, under the umbrella of National Air Communications until March 1940, when they received notice that their jobs no longer existed. Amy's sense of frustration was obvious and her letters to Will reflected this. Aside from the need to earn a living, she wanted to avoid the necessity of having to find other war-related work that she knew would be dull. She drew on her contacts and on her name to secure the right kind of job. In January 1940, the ATA established a Women's Ferry Pool based at Hatfield to take some pressure off the men. The ATA was viewed disparagingly by the RAF, who claimed it stood for "Ancient Tattered Airmen". However, Gerald d'Erlanger's ATA was fulfilling a vital service, and an additional section to harness the piloting skills of many women like Amy seemed an obvious conclusion.

Amy would have been an obvious choice to head the female section, but d'Erlanger favoured Pauline Gower. Amy knew Pauline well from her Stag Lane days, and had kept in touch with her and Dorothy Spicer over the years, occasionally staying at Pauline's home in Kent. Pauline Gower had great respect for Amy's many record-breaking achievements and she wanted her on the ATA team, but the budget was meagre and she'd been restricted to a small number of recruits. Amy was genuinely pleased for her friend, but still felt somewhat peeved to have been overlooked. She'd had reservations about joining the Women's Ferry Pool right up until the last minute, having always seen it as a last resort.

Among the many women who joined their section of the ATA, the original intake was just eight, but the numbers swelled as demand increased. Amy was undoubtedly the most famous, but the recruits came from many diverse backgrounds. Lettice Curtis had graduated from Oxford; Margaret Fairweather was the daughter of a Peer; there was a glamorous blond debutante who had dazzled the London 'Season', modelled for *Tatler* and captained the women's skiing team at the Winter Olympics. Her name was Audrey Sale-Barker, who became Countess of Selkirk. Late in 1932, just after returning from setting a new record to the Cape, she had set off with co-pilot Joan Page. When she joined ATA in June 1940, in a dash of individual flair, she had her uniform made by a Saville Row tailor, the lining of which was bright red silk.

Amy kept the idea of joining the ATA on ice while she pursued other avenues to find work. But in May 1940, she finally succumbed to her friend Pauline's gentle persuasion and joined the team at Hatfield. While the likes of Audrey Sale-Barker gave the ATA its glamour, Amy Johnson gave it gravitas.

The war news was bleak. Norway fell to the Germans that May, along with Belgium and Holland. Neville Chamberlain resigned as Prime Minister, but nobody was sorry to him go. His appeasement of Hitler at Munich had bought Britain time, but nothing more. His replacement was the man who had been warning Chamberlain's predecessor Stanley Baldwin about the potential threat of German rearmament for years, but his had been a lonely voice in the wind. Winston Churchill had his faults and his critics, but he also had guts, the inspirational oratory and an unshakable self-belief that it was his destiny to see Britain through her darkest hour.

Amy took time to adjust to life in the ATA. She attempted to explain her reasons for this in a letter to her mother, saying that she was too much her own woman. Whether she was still smarting from losing out to Pauline Gower for the senior officer's job or not is unclear, but she had little time in which to dwell on personal problems or to get bored. In June, France fell to the Germans within a week and Hitler was merciless in his humiliation of them, insisting that the French sign a surrender document in the same railway carriage that the Germans had formally surrendered on Armistice Day in 1918. The British Expeditionary Force retreated, and in June the largest evacuation ever took place at Dunkirk. Britain now stood alone. In the long, hot, sultry summer of 1940, only the courage and resilience of the RAF in the skies over southern England, saved Britain from imminent invasion.

Amy was far from pleased to learn that Jim Mollison had joined the ATA in October and was stationed at White Waltham. There was much speculation in the mess about what might happen when he bumped into his former wife. Amy was far from happy that Jim tried to seek her out whenever she was at the White Waltham base and complained about him in a letter to Will. Jim's second marriage to Phyllis Hussey had stalled somewhat, as she was independently wealthy and had no need to earn a living. Consequently, she enjoyed her leisure time lounging on a Caribbean beach while Jim preferred to spend his propping up the bar. Jim's admission into the ATA raised a few eyebrows, but his piloting skills were never questioned. He'd been trained by the RAF and, during his five-year service, had seen action on India's NWFP. However, his behaviour in the mess bar had got him into strife then and many believed it might do so again.

Amy had settled into a pattern and seemed to have reconciled herself to her lot in the ATA until the war was over. As the dark weeks of autumn slid into winter and Christmas approached, there seemed little worth celebrating. The war news remained bleak and food shortages were taking their toll. Stoicism still prevailed and, with it, a determination to remain defiant, but every major city had been affected by the Blitz. Amy had not been home since July and was hoping to spend the festive season with her parents who had decided to remain in "Brid". The last of their offspring had flown the nest, as Betty Johnson, now aged twenty-one, had recently got engaged to a young Scotsman named Ronald Falconer-Stewart. She was doing her bit for the war effort, having joined the ATA as a typist.

Instead, Amy spent Christmas 1940 stranded by bad weather in Prestwick. She vented her feelings in a letter home on Boxing Day. The Orangefield Hotel was fine for a night's stopover, but not for much more than that. Amy was upset that the bad weather had robbed her of the chance to see Mollie's baby, the first Johnson grandchild. It was something she had never been able to give her parents, although she vowed to play the role of doting aunt whenever she got chance.

Amy's frustration at waiting for the weather to turn finally ran out and she returned to Hatfield by train. It was a wretched experience. The trains were overcrowded and Amy had to put up with no seat and being constantly recognised by fellow passengers. She much preferred the privacy and freedom of her cockpit. She was back at Hatfield for New Year, but celebrations were muted. There was little appetite for ushering in a New Year with prospects as bleak as the one it was replacing.

There was, however, to be a party on 6th January 1941 to mark the first anniversary of the Women's Ferry Pool and Amy wanted to be there. Duty called in the meantime, however, and on Friday 3rd January, she was given orders to deliver an Airspeed Oxford to Prestwick and bring another back to RAF Kidlington. She had done the Hatfield to Prestwick run so frequently now that it was becoming mundane. Amy climbed into the cockpit, gave a quick wave to her colleagues on the tarmac, hoping to see them at the bash. They, too, were looking forward to the chance to kick up their heels and, for one night at least, forget the gloomy war news. They hoped that Amy would be there to share it with them. Sadly, they never saw her again.

18

THE LAST FAREWELL

The weather was awful from the moment Amy took off from Hatfield, and grew steadily worse until she decided to land at an airfield near Birmingham. She spent the night at a swanky hotel with its own eighteen-hole golf course at Weston-under-Redcastle in Shropshire. The Hawkstone was the kind of hotel she could have imagined staying in after the Australia flight or during her marriage to Jim, when fellow guests would have swarmed around her like bees, begging for her autograph. In 1941, it was still rather grand, but due to war-time restrictions prohibiting it from offering the luxuries its discerning clientele were used to, it was living off its reputation and former glory.

Amy set off next morning and reached Prestwick. She met an old friend, Jennie Broad, who had joined the ATA after Amy and who was hoping to hitch a ride. A last-minute change of instructions from Pauline Gower sent Jennie to catch a train south, while Amy had delivered one Airspeed Oxford to Prestwick and now had to take another south to RAF Kidlington alone.

With the weather still far from favourable, Amy had indicated that she might stay a night at RAF Squires Gate near

Blackpool and seize the opportunity to spend some time with her sister Mollie, brother-in-law Trevor and young niece Susan. ATA rules prohibited pilots from flying after dusk and Gerald d'Erlanger liked his pilots to adhere to the dictum "You're paid to be safe, not brave". Amy therefore felt she could spend a few hours with her family that had been denied her by foul weather at Christmas, and then she would complete her journey to Kidlington on Sunday.

The journey south from Prestwick to Squires Gate was as horrid as the one from Hatfield the previous day. Amy was not in the best of moods. Aside from the weather, the compass on her Airspeed was causing her problems, and she had to negotiate a difficult course, bypassing the barrage balloons to the west of Liverpool. So she was mightily relieved to land in the gloom of a winter's afternoon at the aerodrome near Blackpool and rest there with her family for the night.

Amy and Mollie had much to catch up on. They had been greatly concerned, since the war had started, about their parents and Amy had suggested that they and Mollie escape to Canada while they still had the chance. However, 'Ciss' had been reluctant and the idea was abandoned. She also brought news of their youngest sibling Betty, who was finding her feet in the typing pool of the ATA. Next morning dawned and the weather was still foul. Mollie was eager to dissuade her sister from flying, but Amy was inclined to set out, fly above the clouds and "smell" her way through. She had shared her concerns about the reliability of her compass and this heightened their sense of anxiety, thus urging her to wait for better weather.

When Amy arrived at Squires Gate, the mist was low and dense. The advice was that she shouldn't risk it. She sat in her Airspeed Oxford with Harry Banks, a fueller, and enjoyed a chat and a cigarette. The Airspeed tanks were filled with over 150 gallons, for a journey that should take her approximately an hour and half at the most. Despite a last plea from the duty pilot, who strongly advised her against it, Amy was still determined to give it a go, even though the visibility was poor and Amy had always been superstitious about flying on Sundays. Although there were countless rumours after the events of January 1941 that Amy was on some secret mission and flying with a mysterious 'Mr X' as passenger, although the ATA rules strictly prohibited it, everyone on the ground at Squires Gate unanimously said that Amy took off alone at eleven minutes to midday. Every opinion offered to her on the tarmac like that from her sister Mollie, who had waved her goodbye urged her to wait. The final decision, however, was Amy's only to make. It was a decision that defined her destiny.

The theory held by many who knew Amy when seeking an explanation as to why she was so determined to take off in such foul weather, was her reputation and pride. This may also account for the fact that she never turned back. By three thirty in the afternoon, the light was fading and a quick glance at the fuel gauge told her she was running desperately low. However, the cloud mass was still as dense and Amy was struggling to find a gap. The Airspeed's yellow underbelly would have told anti-aircraft batteries littered along the Thames Estuary that the plane was 'friendly' but only if they could see it in time. Amy spent ages watching for a gap and now she was over two hours' late.

Convoy CE21 consisted of seventeen vessels, among them HMS Haselmere, whose C.O. was Lieutenant Commander Walter Fletcher. At thirty-four, he was already experienced beyond his years.

He had followed in Admundsen's wake across the Northwest Passage to Canada, and on 5th January 1941, due to a selfless act of courage on his part, he was to have his fate entwined for eternity with Amy Johnson. A woman whose name he would have doubtless been familiar, and who he might easily have met, as Fletcher's widowed mother lived at Princes Risborough, the same picturesque village in Buckinghamshire that Amy had lived in during the "wilderness" years after her divorce.

When Amy realised her fuel was virtually gone, she knew that she had to bail out. She had never abandoned her aircraft before and the prospect of a parachute jump scared her. This time, she had no other choice. She jettisoned her aircraft and began floating downwards to the water. At the right moment, Amy pulled the cord and hoped. As she saw the watery depths into which she was going to plunge, a new fear gripped her.

As the icy waters swallowed her up, Amy shouted, "Hurry! Please Hurry!" to anyone who might hear. Ironically, these were the same words she had uttered to rescuers in the inky blackness of Bridgeport, Connecticut, the night *Seafarer* crashed just over seven years before. It was not until Amy uttered those words that Henry O'Dea, who on Fletcher's orders had put the motor launch to sea, realised that the pilot they were trying to rescue might actually be a woman. When O'Dea and another seaman named Dean realised this, they threw out lines to catch her, but the rising swell of the water pushed them further away from Amy's struggling grasp. The icy coldness of the water made her weaker and a sense of panic set in. One of the seamen reached over to lie on the outer rubber ridge of the launch, and stretched as far as he could to reach Amy's outstretched hand. Sadly, Amy was pulled away from her rescuers, and when the swell rose again, the stern of the ship lifted and fell, crashing down onto her. Amy was tragically caught, beyond help, in the propeller blades.

Amid the sense of urgency to help the two figures they believed they had seen emerging from the cloud-laden skies, there was much confusion. Amy's pig-skin bag, and even the door of the Airspeed Oxford, were mistaken for people in the dwindling light of a winter's afternoon. Once they had seen the female pilot vanish from view, Walter Fletcher ordered the two seamen back, while he plunged into the water to rescue what he believed to be a second body. By the time he realised, it was hopeless. He had been in the water too long and was pulled back onto the launch unconscious. Sadly, he never regained consciousness. Walter Fletcher died later that night in the Royal Naval Hospital in Gillingham from his exposure to the icy waters of the Thames Estuary. His was a tragic and unnecessary death, the result of a truly selfless act of courage.

As if to compound their misery, the convoy came under attack from German planes shortly afterwards and *HMS Berkeley* returned anti-aircraft fire. When calm was restored, they went in search of survivors. They managed to retrieve Amy's pig-skin bag, as well as a second bag from the wreckage of the Airspeed Oxford. This revealed the identity of the pilot the *Haslemere* crew had tried so valiantly to save: Amy Johnson. This, along with her log book, was retrieved from the icy depths that claimed her.

When Pauline Gower received the news that Amy's papers had been retrieved from the Thames Estuary and that a plane had crashed that afternoon, she began a frantic sweep of all aerodromes seeking confirmation that Amy had landed. None came. As the reality of tragedy became apparent, duty took over. A telegram was sent to Will and 'Ciss' in Bridlington with a stark message: *Missing believed killed.* In a gesture of courtesy, she telephoned them to soften the blow. It was the news they had dreaded receiving many times over the years, and that they braced themselves for, over a decade before, when Amy was feared missing over Attombea en route to Australia. How ironic, then, that it should finally come, even in the uncertainty of war, when Amy was engaged in routine ferry work over home waters. Will and 'Ciss' bore their loss with bravery and fortitude. They had now endured the experience of losing a child twice.

The idea that Amy Johnson, the Lone Girl Flyer who the nation had taken to its heart when she flew to Australia, and had become a role model to a generation of women, could be lost and feared drowned in the Thames Estuary was hard for the public to accept. They demanded an ending fit for a heroine, so speculation that Amy was on some kind of secret mission for the Allies took on a life of its own. The Admiralty inadvertently helped the rumour mill gather momentum by claiming there had been two bodies in the water: Amy's and the mysterious Mr 'X', the passenger she was supposedly carrying. However heroic and well intentioned these stories were, they served to damage Amy's reputation by suggesting she could have disobeyed ATA rules so flagrantly, causing considerable hurt and embarrassment to her grieving parents.

Her old friend William Courtenay attempted to mend some of the damage by drafting an article that showed Amy in the best possible light. She had used her contacts in the early months of the war to secure a meeting with Lord Vanisttart, who had been instrumental in the establishment of the Special Operations Executive. He had been of the opinion that Amy was not suited to espionage work. She was simply too famous, too recognisable, to have succeeded in covert operations. Consequently, he had had to turn her down.

The grieving process was difficult enough for her parents and continuing rumours about Mr 'X' did nothing to help. Without her body, there could be no funeral service – a crucial part of the grieving process, which they were sadly always denied. There was a memorial service held nine days after on 14th January at St Martin-in-the-Fields. Will, 'Ciss' and their daughter, Betty, attended, along with Amy's friends, John and Alice Hofer, whom she had known for many years and who she had lodged with at their home in Woburn Green.

The service was also attended by many dignitaries and her ATA colleagues, who felt very strongly the words expressed by Pauline Gower in her letter to Will and 'Ciss' that in Amy they had all lost a great friend.

Before making the homeward journey north, Will and 'Ciss'

visited the offices of Charles Crocker & Sons, Amy's former employers, who she had trusted to handle her legal affairs ever since. Now her parents would place the same trust in them to establish the facts about Amy's death and to secure the Grant of Probate that would allow them to formally wind up her estate. It took nearly three years before the seamen serving on HMS *Haselmere* were available to appear in court. Crocker complained on more than one occasion over the intervening years that the Admiralty had, to put it mildly, been less than co-operative. Although the facts had become blurred with time, there was sufficient evidence to grant a 'Presumption of Death', which, due to the war, received scant coverage in the press.

In his memoirs, *Far from Humdrum*, Charles Crocker mentioned that Amy's pig-skin flight bag, which was notoriously mistaken for the helmet of the mysterious Mr 'X', had sat in his office for over a year.

It would later be one of Amy's many belongings that her parents donated to the Museum at Sewerby Hall, near Bridlington, which Amy had officially opened in June 1936 in the glorious afterglow of her second Cape Town record.

Jim Mollison took Amy's death particularly hard. In the three months since he had joined the ATA and was stationed at White Waltham, he had tried to forge an understanding with his former wife, despite there being little encouragement from Amy in that regard – which was understandable, given the pain that Jim had caused her, often very publicly, after their final separation.

He never gave up, however, and Amy was genuinely surprised to receive the new pair of slippers that he bought her for Christmas. He always kept his true feelings for Amy strictly private, and in his grief, he took refuge in alcohol. His view on the circumstances of her death, however, was unequivocal. Jim believed that Amy was shot down by enemy fire on that grey Sunday afternoon in January 1941. He had no firm evidence to support his theory, but he had his own experience of having been attacked. He believed that Amy had suffered the same fate until the end of his days.

Fate can sometimes be mercilessly cruel and the Johnson family suffered many heartaches over the years. They were to suffer another blow in January 1944 when Betty's pilot husband, Ronald Falconer-Stewart DFC, was killed in action, along with five members of his Lancaster bomber crew, leaving Betty like many of her generation, a widow, at the age of twenty-three. The two granddaughters that Will and 'Ciss' gained through their daughter Mollie were a great joy and a constant source of comfort to them in their final years. Amy remained in their thoughts and one room of the house in Beverley was turned over to the many photographs and souvenirs of her illustrious flying career. While concerns over Amy's safety had always been uppermost in their minds when she was on her record-breaking flights, they were flushed with pride by all that their daughter had achieved in aviation. Will, who had enjoyed a special bond with Amy, felt her loss very keenly at times.

Amy Johnson's legend, due in some part to the mysterious circumstances surrounding her final hours, has become embalmed in history for over seventy years. As Amy herself once said:

"I have always believed that those who live colourfully, fight courageously and love greatly, remain to enrich our memories and shape our actions. Such is real fame."

Amy managed, in her relatively short life, to do all of those things and so much more.

EPILOGUE

Perhaps it is too simplistic to say that Amy Johnson broke the rules on that fateful day, but by waving aside the advice of those on the ground at Squires Gate, she probably did. While the Air Transport Auxiliary had rules about taking off in bad weather, the final decision was always left to the pilot.

As an experienced flyer, Amy believed she could still make it to RAF Kidlington by going over the top and 'smelling' her way through. Sadly, not this time and so she died in the active service of her country in a time of war. She was one of the many brave men and women who, in countless different ways, made the ultimate sacrifice.

It is, however, strangely ironic that the Lone Girl Flyer who braved the vast shark-infested waters of the Timor Sea on the last leg of that epic flight should end her life in the dark icy waters of the Thames Estuary just as it had begun beside another estuary one golden summer thirty-eight years before. The years between these events mark the unfolding story of a truly remarkable woman, born with the lust for adventure. The cigarette-card heroine whose image is frozen in our memories as a fresh-faced beauty in a helmet and goggles, primed for action.

In 1930, Australia represented the greatest challenge to any pilot with the courage and ambition to conquer the sky lanes of the world. Amy had always dreamt big, but she also possessed the determination to make her dreams reality. So, when she brought her battered Gypsy Moth biplane down onto the pot-holed runway of Darwin's Fanny Bay aerodrome, after nineteen gruelling days in the air, not only did she succeed in bringing the far-flung people of the Empire closer together, she earned her place in the annals of aviation history and the lasting affection of a grateful nation.

ACKNOWLEDGEMENTS

This book could not have been written without the assistance and support of number of people.

Firstly I would like to thank staff at Crawley Library in West Sussex, who ordered in most of the titles listed in my select bibliography. My thanks also go to staff at the Central Library, Hull. I made two visits to the city of Amy's birth, and on the first in August 2004, I was given access to the extensive microfilm archive of over 300 letters, that Amy had written to Hans Arregger. The Local Studies Library in Katherine Street, Croydon were also generous in the time they gave me to trawl through their archive of Amy Johnson related material.

The Department of Research and Information Services at the RAF Museum in Hendon holds an extensive Amy Johnson archive, and they too were enormously generous in the time they gave me to read some of Amy's personal correspondence and look through photographic images. Dr David Marchant, Museums Registrar at East Riding of Yorkshire Council, gave up a large chunk of his time showing me around the Amy Johnson Room at Sewerby Hall, near Bridlington. They hold many of the gifts and souvenirs given to Amy on her pioneering flights.

Vanna Skelley at the Castrol Archive was an valuable source of help in my quest for information about the relationship between Amy and Lord Wakefield of Hythe, Chairman of Castrol Oils. I am also grateful for their assistance to the volunteers of the Croydon Airport Society, who I 'pumped' for information about Amy, and Croydon Airport in general as the Airport of Empire was the starting point of so many of Amy's heroic flights.

To fellow biographer Midge Gillies, I would like to express my thanks for your assistance and encouragement, as my first correspondence to you came so soon after your book had been published. *Queen of the Air* has been widely lauded as the 'definitive' biography of Amy Johnson, so for me you're a hard act to follow.

To my family, and friends, thank you all for your patience and support over the years it has taken me to write this book.

SELECT BIBLIOGRAPHY

Babington Smith, Constance, *Amy Johnson*, Collins 1967

Christie, Agatha, *Peril at End House,* Collins 1932

Courtenay, William, *Airman Friday*, Hutchinson 1937

Crocker, Charles, *Far from Humdrum*, Hutchinson 1967

Dixon, Charles, *Amy Johnson* – Sampson 1930

Fox, James, *White Mischief,* Jonathan Cape 1982

Gillis, Midge, *Amy Johnson: Queen of the Air,* Weidenfeld & Nicolson 2003

Higham, Charles, *Howard Hughes: The Secret Life,* Virgin Books 2004

Johnson, Amy, *Myself when Young*, Frederick Muller 1938

Johnson, Amy, *Sky roads of the World*, Chambers 1939

Luff, David, *Amy Johnson, Enigma in the Sky*, Airlife Publishing 2002

Luff, David, *Mollison: The Flying Scotsman*, Lidun Publishing 1993

Mackersey, Ian, *Jean Batten: The Garbo of the Skies*, Macdonald 1991

Mollison, James, *Death Cometh Soon or Late*, Hutchinson 1932

Mollison, James, *Playboy of the Air*, Michael Joseph 1937

Nesbit, Roy, *Missing, Believed Killed*, Sutton Publishing 2002

Rich, Doris L, *Amelia Earhart: A Biography*, Air life Publishing 1989

Trzebinski, Errol, *The Lives of Beryl Markham*, Norton 1995

Whittell, Giles, *Spitfire Women of World War II*, Harper Perennial 2008

Television, *the Real Amy Johnson*, Lion Television 2003

Film, *They Flew Alone*, RKO Radio Pictures 1941

RESEARCH NOTES

For the bulk of my research I relied on the previous biographies of Amy Johnson, which throughout these research notes will be listed in chronological order: *Amy Johnson* by Constance Babington Smith, *Enigma in the Sky* by David Luff, and *Queen of the Air* by Midge Gillies. The other sixteen titles listed in the bibliography also helped me to construct this portrayal, of one of the most truly remarkable women of the 20th century.

1. (THE FISH-MERCHANT'S DAUGHTER)
AMY JOHNSON, CBS
ENIGMA IN THE SKY
MYSELF WHEN YOUNG

2. (AMY JOHNSON: BACHELOR OF ARTS)
AMY JOHNSON, CBS
ENIGMA IN THE SKY
SELECTED LETTERS TO HANS ARREGGAR HCL

3. (HANS ARRANGER: HER ERRANT KNIGHT)
AMY JOHNSON, CBS

ENIGMA IN THE SKY
MYSELF WHEN YOUNG
SELECTED LETTERS TO HANS ARREGGAR HCL

4. (EXILE IN LONDON)

AMY JOHNSON, CBS
ENIGMA IN THE SKY
MYSELF WHEN YOUNG

5. (EARNING HER WINGS)

AMY JOHNSON, CBS
ENIGMA IN THE SKY
AMELIA EARHART: A BIOGRAPHY
JEAN BATTEN: THE GARBO OF THE SKIES

6. (AUSTRALIA: THE GEM OF AN IDEA)

AMY JOHNSON, CBS
ENIGMA IN THE SKY
SKYROADS OF THE WORLD

7. (CROYDON TO KARACHI)

AMY JOHNSON, CBS
ENIGMA IN THE SKY

8. (THE LONE-GIRL FLYER)

AMY JOHNSON, CBS
ENIGMA IN THE SKY
QUEEN OF THE AIR

9. (AUSTRALIA!! A WHIRLWIND TOUR)

AMY JOHNSON, CBS
LONE-GIRL FLYER
DEATH COMETH SOON OR LATE

10. (CELEBRITY STATUS: AT HOME & ABROAD)
QUEEN OF THE AIR
AIRMAN FRIDAY
LONE-GIRL FLYER

11. (TO THE LAND OF THE RISING SUN)
ENIGMA IN THE SKY
QUEEN OF THE AIR
MOLLISON: THE FLYING SCOTSMAN

12. (AMY & JIM)
MOLLISON: THE FLYING SCOTSMAN
DEATH COMETH SOONER OR LATE
PLAYBOY OF THE AIR
AIRMAN FRIDAY
PERIL AT END HOUSE

13. (SOLO TO THE CAPE)
AMY JOHNSON, CBS
QUEEN OF THE AIR
AIRMAN FRIDAY

14. (DISASTER AT BRIDGEPORT)
ENIGMA IN THE SKY
QUEEN OF THE AIR
MOLLISON: THE FLYING SCOTSMAN
HOWARD HUGHES: THE SECRET LIFE
AMELIA EARHART: A BIOGRAPHY

15. (THEIR LAST HURRAH!)
QUEEN OF THE AIR
THE LIVES OF BERYL MARKHAM
PLAYBOY OF THE AIR
WHITE MISCHIEF

16. (THE WILDERNESS YEARS)
QUEEN OF THE AIR
PLAYBOY OF THE AIR
MOLLISON: THE FLYING SCOTSMAN
SKYROADS OF THE WORLD
LETTER FROM AMY TO DUCHESS OF
BEDFORD, RAFM

17. (THE ATA: A NATION AT WAR)
AMY JOHNSON, CBS
ENIGMA IN THE SKY
QUEEN OF THE AIR
MOLLISON: THE FLYING SCOTSMAN
MISSING: BELIEVED KILLED
SPITFIRE WOMEN OF WORLD WAR II

18. (THE LAST FAREWELL)
ENIGMA IN THE SKY
QUEEN OF THE AIR
MOLLISON: THE FLYING SCOTSMAN
FAR FROM HUMDRUM
THE REAL AMY JOHNSON (Lion Television)

Key CBS = Constance Babington Smith
RAFM = Royal Air Force Museum
HCL = Hull Central Library